T0212393

Springer Texts in Business and Economics

Springer Texts in Business and Economics (STBE) delivers high-quality instructional content for undergraduates and graduates in all areas of Business/Management Science and Economics. The series is comprised of self-contained books with a broad and comprehensive coverage that are suitable for class as well as for individual self-study. All texts are authored by established experts in their fields and offer a solid methodological background, often accompanied by problems and exercises.

More information about this series at http://www.springer.com/series/10099

Pierre Brugière

Quantitative Portfolio Management

with Applications in Python

 Springer

Pierre Brugière (ID)
CEREMADE
University Paris Dauphine-PSL
Paris, France

ISSN 2192-4333   ISSN 2192-4341   (electronic)
Springer Texts in Business and Economics
ISBN 978-3-030-37742-7   ISBN 978-3-030-37740-3   (eBook)
https://doi.org/10.1007/978-3-030-37740-3

Mathematics Subject Classification (2010): 91G10, 91G70

This Springer imprint is published by the registered company Springer Nature Switzerland AG.
The registered company address is: Gewerbestrasse 11, 6330 Cham, Switzerland

Preface

This book presents some well-established quantitative methods for modelling the performance of financial assets, deriving from there optimal portfolios and estimating the parameters of the models used. The hypothesis of normality, for the asset returns, and some statistical tests of this assumption are presented in the first chapter.

In the second chapter, the theory of utility functions is presented to model, for an investor, the problem of choice between different investment strategies. In this chapter, the notions of risk aversion and risk premium are explained, in terms of concavity of some utility functions. From this chapter on, every investment strategy is analysed from a risk/return perspective, the risk being defined as the standard deviation of the returns. This risk/return approach, in a Gaussian law context, has been the cornerstone of an extensive financial literature, starting in the 50s. Harry Markowitz, William F. Sharpe, Jack Treynor, John Lintner, and Jan Mossin are some of the main creators and developers of this theory, whose impact on quantitative asset management has been recognised through a Nobel Prize in Economics for Harry Markowitz, William F. Sharpe and Merton H. Miller in 1990. Even if the assumptions of the model do not always perfectly match the reality, at least, they offer a well-posed framework, under which interesting concepts can be developed.

In this book, we cover the main results of the theory and in particular the Capital Market Line, the two funds theorem, the Tangent and Market Portfolio and the Security Market Line results. The book aims at demonstrating all the core results, very rigorously with no need for exterior references. The demonstrations are often based on simple algebra, and most of the results can be interpreted geometrically, in terms of orthogonal projections or minimisation of a norm, for a geometry defined by the variance-covariance matrix of the risky assets returns. The concept of self-financing portfolios and of their returns is used extensively and eases considerably the description of the construction and of the performance of some efficient investment portfolios. It also offers a bridge with the usual ways of thinking of an asset manager or a trader.

In Chap. 7, we study risk measures, which were first developed in the 1990s and which are now widely used for regulation and risk monitoring purposes under the names of Value at Risk and Expected Shortfall. We link risk measure analysis to the problem of portfolio optimisation and give some examples of calculations both in the Gaussian framework and in the absence of any model assumptions. We also discuss the Euler formula and show how it can be applied to the problem of optimal capital allocation.

The last sections of the book go deeper into some possible statistical specifications of the model, with factor models and the Asset Pricing Theory (APT). Very often, factor models and APT models are introduced from an economic/econometric perspective and the factors consist of variables such as inflation, interest rates, FX, financial characteristics of the companies studied...etc. In such approaches, the factors are therefore exogenous and are brought in because of some prior understanding of how the economy and the financial markets work. A very important literature exists on this subject, aiming at identifying these factors and at quantifying how they explain the returns and the risks of all the stocks. Eugene F. Fama, Lars Peter Hansen and Robert J. Shiller received a Nobel Prize in Economics in 2013 for, amongst other things, their empirical works on the topic. In this book, we will not dive into the subject of searching for exogenous factors, which is an on-going topic of discussion, and which is in fact very market and time-dependent, but we will go to the bottom of the model used and of the statistical hypothesis made, by showing how endogenous factors can be identified via Principal Component Analysis. In this statistical approach, the search for endogenous factors, to explain the risks and the correlations, comes naturally after the Security Market Line result. As a matter of fact, as explained in Chap. 8, the Tangent Portfolio is itself an endogenous factor, and the Security Market Line equation can be seen, mathematically, as a one-factor model explaining the expected returns of all the stocks in the best possible way. This being said, as the Tangent Portfolio explains, only very imperfectly, the risks and the correlations, it appears natural to search for additional statistical factors, which will be designed specifically for this purpose, and this is what statistical factor analysis and Principal Component Analysis do.

As a special case of factor models, APT models are introduced in Chap. 8. Here they are defined as the mathematical solution of a non-arbitrage problem, whereas most of the time in the literature they are presented only as the consequence of a rather vague argument according to which non-factorial risks can be diversified (as much as wanted) and therefore should not be remunerated. Finally, some mathematical properties of APT models are analysed and some important characteristics of these models are demonstrated. Amongst them, the link between the expected return of an asset and the correlation of its risks with the factors.

Some examples are given in the last chapters on how to decompose the risk, identify the statistical factors and conduct a Principal Component Analysis.

The aim of this book is to explain the theory of asset allocation in a risk/return framework, as well as the mathematics of factor models and risk measures, in a very synthetic but rigorous way, with elegant and concise proofs. Only in the first chapter are some proofs skipped, as they use complex convergence theorems in probabilities, which cannot be demonstrated from scratch in a reasonable number of pages. The main results are illustrated and expressed in plain language, understandable without any mathematical background. A line is drawn between the mathematical results derived purely from the model and their usual economic interpretations, which require some extra economic assumptions. Because of this, we make a distinction, for example, between the Tangent Portfolio and the Market Portfolio, as the latter can be identified with the former only at the cost of some extra economic assumptions, related to the behaviours of all the agents in the economy. The book is centred around the theory and its mathematical consequences, which translate into important investment paradigms. The reader with a reasonable background in mathematics should, after reading the book, be able to implement most of the results presented to any market of his choice. The main mathematical prerequisites necessary to understand the demonstrations in this book are related to basic linear algebra, orthogonal projections and random Gaussian variables. These notions should be known to most advanced undergraduate students in mathematics or professionals in a quantitative role. Somebody without a strong quantitative background can skip the mathematical demonstrations and still find an understandable and informative description of the results in plain English. There is also a certain parallel between the theory presented in this book and the Black and Scholes formula. In both cases, everybody will agree that the model is imperfect, but that at the same time it is necessary to understand how things works in these simplified models before considering anything else more complex.

The Python code included enables to extract financial data from the net and to put into practice the methods described. At the time of printing of this book everything was working on Python 3.6 with the standard libraries installed. Simply copy pasting the code into a Jupyter notebook should enable the reader to run all the examples provided. The programs can also be run in the Cloud with Google Colab and the Python 3.6 version it provides. Google Colab is free and can be accessed by creating a gmail address and opening the Google drive associated. The advantage of using Google Colab is that there is no requirement to install Python on one's computer as all calculations are made in the cloud, with the calculation capacity provided by Google. This being said, Python and its libraries, as well as the way data are structured and offered by providers, are evolving quickly. So, inevitably some adjustments will have to be made in the future for the code. If some data extraction issues occur in the future, our advice is to first check if the data provider has changed the ticker of some particular stocks. Finally, the choice of the DAX index, for most examples, is motivated by the fact that it is one of the few major national indices which is calculated on a total return basis, but the reader can adjust the tickers in the program to analyse other indices.

This book comes as the result of a twenty one hour course that I have been giving on the topic to first year Masters students, at Université Paris-Dauphine, since 2015. In the section "Exercises and Problems" figure most of the mid-term and final exams that I produced during this period, with the solutions. Any student following a similar course, or professional interested in a little challenge, can probably train on these exams or at least on the multiple part questions they contain. Readers with a good mathematical background will notice that, very often, the exams are about an important result from the course but treated from a complementary angle. Note that twenty one hours of lectures is in fact quite a short amount of time to cover all the material of the book and somebody in a hurry to grasp the essence of portfolio optimisation, in a mean variance framework, could start directly at Chap. 3, and leave Chaps. 1 and 2 for later readings.

To produce this book I benefited greatly from the lectures notes produced by my predecessors: Idris Kharroubi, Imen Ben Tahar, Anthony Reveillac and Christian Lopez; their notes were my starting point to develop this book. This book has also benefited from discussions and support I received from colleagues from the MIDO and from the Cérémade departments of Université Paris-Dauphine. Among them Pierre Cardaliaguet, Danielle Florens, Gabriel Turinici, Emmanuel Lepinette, Marc Hoffman and Bruno Bouchard, under the impulsion of whom we developed with Pierre Clauss an Executive Master of Asset Management at Université Paris-Dauphine. I would also like to thank Elyès Jouini and Aïda Hamdi and the House of Finance of Dauphine with whom several events were organised at the University related to the topic. This book was also influenced by some experience I acquired while working in investment banking in London for almost two decades until 2015, and by some interesting discussions and exchanges of points of view I had with some former colleagues there. Among them Pierre, Gerald, François, Remi, Alireza, Eric, Gwenael and Laurent will recognise themselves.

I would also like to thank Chloé Rabanel for attentively reading the manuscript and helping me to put it into its final form.

Last but not least, I thank my whole family and my wife and children, for their support while writing this book.

Paris, France   Pierre Brugière

Contents

Returns and the Gaussian Hypothesis

1

In this book, the problem of finding optimal portfolios is mathematically solved under the assumption that the returns of the risky assets follow a Gaussian distribution. In this section, we give the definition of a price return and of a total return and describe some tools to analyse these returns and to statistically test the hypothesis of normality on them. The hypothesis does not always appear to be satisfied, depending on the stock or on the period considered, nevertheless, even in these cases, the methods of portfolio optimisation may still teach some useful lessons.

1.1 Measure of the Performance

1.1.1 Return

We consider an economy with two instants of observations 0 and T. The investment decisions are made at time 0, which is today, and the result of the investment is observed at a future time T. The notion of return is defined as follows:

- The **return of an investment** is defined as $R_T = \frac{W_T}{W_0} - 1$, where W_0 is the initial amount invested and W_T is the value of the investment at time T. W_0 can be seen as a fund initial value or mark to market at time 0 and W_T its final value.
- The **return of an asset** is defined as the return of an investment in this asset. It is necessary when calculating a return to take into account all coupons (for a bond) or dividends (for a share) paid by the asset to its holder during the period of detention $[0, T]$. If P_0 is the value of the asset at time 0, P_T its value at time T and $CR(0, T)$ the cumulated values of all payments (coupons or dividends) made by the asset during the period $[0, T]$ and reinvested from their dates of distributions to time T, we get: $R_T = \frac{P_T + CR(0,T)}{P_0} - 1$. Note that this return is called the **total return** of the asset, while the return calculated without taking into account the term $CR(0, T)$ is called the **price return**.

© Springer Nature Switzerland AG 2020
P. Brugière, *Quantitative Portfolio Management*, Springer Texts in Business and Economics, https://doi.org/10.1007/978-3-030-37740-3_1

It is important, when comparing investments or indices, to compare their total returns, otherwise the conclusions may be erroneous. For example, the French CAC 40 index and the S&P 500 index do not integrate the dividends paid by their constituents into their calculations, and therefore are price return indices, while the German DAX 30 index is a total return index because the dividends paid by its constituents are added to the value of the index. So, comparing the evolutions of indices can be misleading if dividends are reintegrated in some calculations but not in others.

1.1.2 Rate of Return

To the notion of return is associated the notion of **rate of return**. Depending on the instrument or market considered, the definition of the rate may vary: usually for fixed income instruments of maturity less than 1 year at the time of issuance (Treasury bills, short term Commercial Papers) the **monetary rate** is used while for fixed income instruments of maturity above 1 year at the time of issuance (Treasury Notes, Bonds, Medium Term Notes) the **actuarial rate** is used. Finally, in financial mathematical modelling a **continuous rate** is usually used. If T is the measure of time for an investment, these rates are thus defined in the following ways:

- **monetary rate**: $1 + r \times T = 1 + R_T$,
- **actuarial rate**: $(1 + r)^T = 1 + R_T$,
- **exponential rate**: $\exp(r \times T) = 1 + R_T$.

In practice, there are different ways to measure the time T between two dates. Even if a calendar year generally corresponds to a T close to 1, the result may differ slightly from 1 depending on the basis used to measure time. The following bases can be used: 30/360, exact/360, exact/365. For example, to measure the time between September 3, 2018 and December 17, 2018 the results are as follows, depending on the convention used to measure time:

- convention 30/360 : $T = (3 \times 30 + (17 - 3))/360 = 104/360 = 0.28889$,
- convention exact/360 : $T = 105/360 = 0.29167$,
- convention exact/365 : $T = 105/365 = 0.28767$.

As the prices for fixed income instruments are often negotiated in yields, it is important before dealing to verify the conventions used, in order for all parties to come up with the same amount to be paid. Also, for some short term instruments the rate used may be a **prepaid rate**, leading to a different formula for the calculation of the dealing price. So, traders should know all these conventions and the instruments and markets which they refer to before dealing. This information is available in specialised documentation (see, for example, Steiner [86]).

Remark 1.1.1 The continuous rates of return, also called exponential rates of return, satisfy the following interesting mathematical properties:

- They compound easily. So, if $r(0, T_1)$ is the continuous rate of return over $[0, T_1]$ and $r(T_1, T_1 + T_2)$ is the continuous rate of return over $[T_1, T_1 + T_2]$ then the continuous rate of return $r(0, T_1 + T_2)$ over $[0, T_1 + T_2]$ satisfies:

$$\exp(r(0, T_1 + T_2)(T_1 + T_2)) = \exp(r(0, T_1)T_1)\exp(r(T_1, T_1 + T_2)T_2)$$

$$\Rightarrow r(0, T_1 + T_2) = \frac{r(0, T_1)T_1 + r(T_1, T_1 + T_2)T_2}{T_1 + T_2}.$$

This property is interesting when doing stochastic modelling as assuming that rates of returns on distinct periods follow independent normal laws will imply that the rate of return for any period will follow a normal law as well.

- When compounding a constant instantaneous rate of return r, the continuous interest rate of return obtained for any period will be this interest rate r. This mathematically translates into

$$\forall t > 0, \, \mathrm{d}P_t = r P_t \mathrm{d}t \Rightarrow \forall T > 0, \, P_T = P_0 e^{rT}.$$

1.2 Probabilistic and Empirical Definitions

Some tests of normality, on a random variable, can be conducted by analysing its moments. This is what is done here, with the calculations of the moments of order 3 and 4, after renormalising the random variable (centering and reducing it) by its mean and standard deviation. The probabilistic definitions of the moments, given below, are applied to a sample by calculating them for the sample **empirical probability**. More generally, when considering a random variable $X : (\Omega, P) \longrightarrow \mathbb{R}$, defined on a probabilistic space (Ω, P), and a measurable function $f : \mathbb{R} \longrightarrow \mathbb{R}$, integrable under P, then, to the probabilistic quantity $\mathbf{E}_P(f(X)) = \mathbf{E}_{P^X}(f(x))$ is associated the empirical quantity, $\mathbf{E}_{\hat{P}^X}(f(x))$, also called the **plug-in estimator**. Here, $\hat{P}^X$ is the empirical probability of X, derived from the sample $(x_i)_{i \in [\![1,n]\!]}$ of X by associating the probability $\frac{1}{n}$ to each of the observations x_i. The probabilistic definitions used in this chapter are as follows.

Definition 1.2.1 (Probabilistic Definitions for a Random Variable X)

- **Expectation**: $\mathbf{E}(X)$,
- **Variance**: $\mathbf{Var}(X) = \mathbf{E}(X^2) - \mathbf{E}(X)^2$,
- **Standard Deviation**: $\sigma(X) = \sqrt{\mathbf{Var}(X)}$,
- **Skew**: $\mathbf{Skew}(X) = \mathbf{E}\left(\left(\frac{X - \mathbf{E}(X)}{\sigma(X)}\right)^3\right)$,

- **Kurtosis**: $\mathbf{Kur}(X) = \mathbf{E}\left(\left(\frac{X - \mathbf{E}(X)}{\sigma(X)}\right)^4\right)$ (some authors subtract 3 from this quantity and call the result the **excess kurtosis**).

To these probabilistic definitions correspond the following empirical definitions, or plug-in estimators, by taking as a particular probability the empirical probability $\hat{P}^X$ derived from a sample.

Definition 1.2.2 (Empirical Definitions for a Sample $x = (x_1, x_2, \cdots x_n)$)

- **Sample Mean**: $\hat{\mathbf{E}}(x) = \frac{1}{n} \sum\limits_{i=1}^{i=n} x_i$ also denoted $\bar{x}$,

- **Sample Variance**: $\widehat{\mathbf{Var}}(x) = \frac{1}{n} \sum\limits_{i=1}^{i=n} (x_i - \bar{x})^2$,

- **Sample Standard Deviation**: $\hat{\sigma}(x) = \sqrt{\widehat{\mathbf{Var}}(x)}$,

- **Sample Skew**: $\widehat{\mathbf{Skew}}(x) = \frac{1}{n} \sum\limits_{i=1}^{i=n} (\frac{x_i - \bar{x}}{\hat{\sigma}(x)})^3$,

- **Sample Kurtosis**: $\widehat{\mathbf{Kur}}(x) = \frac{1}{n} \sum\limits_{i=1}^{i=n} (\frac{x_i - \bar{x}}{\hat{\sigma}(x)})^4$.

We now exhibit some properties that the probabilistic quantities defined above satisfy for a normal law. We write $X \sim Y$ when X and Y follow the same laws. We abbreviate the expression "random variable" by *r.v.* and "independently equidistributed" by *i.i.d.*

Property 1.2.1 (Skewness)

- anti-symmetry: $\mathbf{Skew}(-X) = -\mathbf{Skew}(X)$,
- scale invariance: if $\lambda > 0$, $\mathbf{Skew}(\lambda X) = \mathbf{Skew}(X)$,
- location invariance: $\forall \lambda \in \mathbb{R}$, $\mathbf{Skew}(X + \lambda) = \mathbf{Skew}(X)$,
- if $X \sim N(m, \sigma^2)$ then $\mathbf{Skew}(X) = 0$.

Property 1.2.2 (Kurtosis)

- symmetry: $\mathbf{Kur}(-X) = \mathbf{Kur}(X)$,
- scale invariance: if $\lambda \neq 0$, $\mathbf{Kur}(\lambda X) = \mathbf{Kur}(X)$,
- location invariance: $\forall \lambda \in \mathbb{R}$, $\mathbf{Kur}(X + \lambda) = \mathbf{Kur}(X)$,
- if $X \sim N(m, \sigma^2)$ then $\mathbf{Kur}(X) = 3$.

Proof Easy and left to the reader. To calculate the kurtosis integration by parts is used, which implies that for any integer $n > 1$: $\int_{-\infty}^{+\infty} x^n e^{-\frac{x^2}{2}} \, dx = (n - 1) \int_{-\infty}^{+\infty} x^{n-2} e^{-\frac{x^2}{2}} \, dx$. □

Definition 1.2.3 If $\mathbf{Kur}(X) > 3$, the distribution is said to be **leptokurtic** or to have "fat tails", and if $\mathbf{Kur}(X) < 3$ the distribution is said to be **platykurtic**.

Usually, the argument given against the assumption of normality of the returns is that the kurtosis observed is higher than expected (i.e. fat tails for the empirical distribution).

Definition 1.2.4 The **empirical probability** $\hat{P}^X$ derived from a sample $x = (x_1, x_2, \cdots x_n)$ is defined by

$$\hat{P}^X(\cdot) = \frac{1}{n} \sum_{i=1}^{n} \delta_{x_i}(\cdot),$$

where $\delta_{x_i}(\cdot)$ is the Dirac measure defined by $\delta_{x_i}(x) = 1$ if $x = x_i$ and zero otherwise.

Remark 1.2.1 The properties of symmetry, location invariance and scale invariance are true for the empirical quantities as well, as this is a particular case of the general probabilistic results, when replacing the probability P^X by the probability $\hat{P}^X$.

Remark 1.2.2 For a mixture of normal distributions the kurtosis is > 3. For example, let X be a Bernoulli variable $X \sim \mathcal{B}(0.5)$ and Z_1, Z_2 be two independent normal distributions of laws $\mathcal{N}(m_1, \sigma^2)$ and $\mathcal{N}(m_2, \sigma^2)$. Then the random variable $Z = X Z_1 + (1 - X) Z_2$ is called a mixture of normal variables and it can be shown (left to the reader) that if $m_1 \neq m_2$ then $\mathbf{Kur}(Z) > 3$.

1.3 Goodness of Fit Tests

From the fact that, for a Gaussian variable, $\mathbf{Skew}(X)$ is zero and $\mathbf{Kur}(X)$ is equal to 3, it is expected that, when a distribution is Gaussian, the empirical quantities $\widehat{\mathbf{Skew}}(X)$ and $\widehat{\mathbf{Kur}}(X)$ should converge, as the sample size increases, towards these values. The Bera–Jarque theorem below shows that, if the underlying distribution is Gaussian, there is indeed convergence and that the speed of convergence is of the order $\sqrt{n}$. The theorem also says that the limit normal distributions obtained can be squared and added, as two independent normal distributions could be, to produce a $\chi^2(2)$ distribution. Based on this, the Bera–Jarque test encapsulates in a single number the strength of the hypothesis for both the calculated skew and kurtosis.

Theorem 1.3.1 (Bera–Jarque Test) *Let* $X_1, X_2, \cdots X_n$ *be i.i.d.* $\mathcal{N}(m, \sigma^2)$ *and* $X = (X_1, X_2, \cdots, X_n)$. *Then*

(1) $\sqrt{n}\widehat{\textbf{Skew}}(X) \xrightarrow{Law} \mathcal{N}(0, 6)$,

(2) $\sqrt{n}(\widehat{\textbf{Kur}}(X) - 3) \xrightarrow{Law} \mathcal{N}(0, 24)$.

If $\widehat{BJ}(X) = \frac{n}{6}(\widehat{\textbf{Skew}}(X))^2 + \frac{n}{24}(\widehat{\textbf{Kur}}(X) - 3)^2$, *then* $\widehat{BJ}(X) \xrightarrow{Law} \chi^2(2)$. *If* $\chi^2_\alpha(2)$ *is such that* $P(\chi^2(2) > \chi^2_\alpha(2)) = \alpha$ *then the Bera–Jarque test rejects at confidence level* α *the normality hypothesis iff* $\widehat{BJ}(x) > \chi^2_{1-\alpha}(2)$.

The proof of this theorem can be found in Bera and Jarque [19].

1.3.1 Example: Testing the Normality of the Returns of the DAX 30

Figure 1.1 shows the closing prices of the DAX 30 for the year 2017. The DAX 30 is a total return index, so the dividends distributed by the stocks composing the index are integrated into the calculations of the prices and the returns.

All the daily returns for the year 2017 are represented in Fig. 1.2 and appear to be in the range [−1.83%, 3.37%].

To test the hypothesis of normality for the 251 daily returns of this sample, the following statistics are calculated:

- $\bar{x} = 0.045\%$, $\widehat{\sigma}(x) = 0.667\%$,
- $\widehat{\textbf{Skew}}(x) = 0.55$ and $\widehat{\textbf{Kur}}(x) = 5.54$,
- $\widehat{BJ}(x) = 80.12$ and $\chi^2_{5\%}(2) = 5.99$.

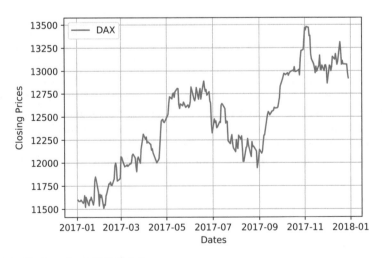

Fig. 1.1 Closing prices, DAX 30 index

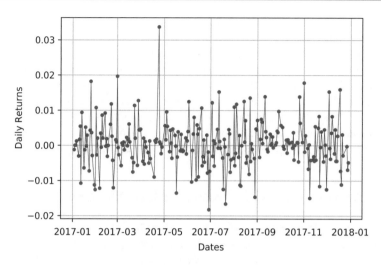

Fig. 1.2 Daily returns, DAX 30

So, according to the Bera–Jarque test the normality hypothesis is rejected at confidence level 95% as 80.12 is not in the 95% confidence interval $[0, 5.99]$. The fact that the observed distribution has a tail fatter than expected (kurtosis of 5.54 instead of 3) and presents some asymmetry (skew of 0.55 instead of 0) is considered here to be a significant deviation from the limit values they should have, for a sample of this size.

Remark 1.3.1 The Bera–Jarque test is very sensitive to outliers. When doing the same test but on the second part of 2017, where the 3.37% spike is not present and with 127 observed returns, we obtain:

$$\frac{n}{6}(\widehat{\mathbf{Skew}}(X))^2 = 0.72 \text{ and } \frac{n}{24}(\widehat{\mathbf{Kur}}(X) - 3)^2 = 1.38$$

and for this period the Skew and the Kurtosis observed are consistent with a normal law assumption and the Bera–Jarque test is satisfied.

Remark 1.3.2 The volatility is defined as $volatility \times \sqrt{\Delta T} = \widehat{\sigma}(x)$, where x is the sample of the log returns and ΔT is the average of the lengths of the periods over which each of the returns is calculated. Here, for daily variations the log returns and the returns are very similar numbers. Also, $\Delta T = \frac{1}{251}$ as there are 251 returns observed in 1 year. So, the estimation of the volatility here is 10.45%.

1.4   Further Statistical Results

An idea to test the normality assumption is to calculate a density function estimate f_n from the sample, to infer the density function f of the variable. The Parzen–Rosenblatt theorem justifies this approach and the use of a histogram to determine the nature of the distribution. Some results are also presented, based on the maximum distance between the empirical cumulative distribution function and the cumulative distribution function of a normal distribution, in the form of the Kolmogorov–Smirnov theorem. This theorem is the basis of many statistical tests.

1.4.1   Convergence of the Density Function Estimate

Theorem 1.4.1 (Parzen and Rosenblatt Estimation of the Density) *Let X be a random variable with density function $f(x)$. Let $(X_i)_{i\in\mathbb{N}}$ be i.i.d. variables with the same law as X. Let K be positive of integral 1 and $(h_n)_{n\in N}$ be such that $h_n \to 0$ and $nh_n \to \infty$. Let $f_n(x)$ be defined by $f_n(x) = \frac{1}{n}\sum_{i=1}^{n}\frac{1}{h_n}K(\frac{X_i-x}{h_n})$. Then under certain regularity conditions, for any x in $\mathbb{R}$ we get,*

$$f_n(x) \text{ is a density and } \sqrt{nh_n}\big(f_n(x) - f(x)\big) \xrightarrow{Law} \mathcal{N}\Big(0, \, f(x)\int_{-\infty}^{+\infty}K^2(x)\mathrm{d}x\Big).$$

For a proof, see Parzen–Rosenblatt [69].

Example 1.4.1 The following functions are often used as kernels:

- rectangular kernel: $K(u) = \frac{1}{2}1_{|u|<1}$ or more generally,
- rectangular kernel of window h: $K_h(u) = \frac{1}{2h}1_{|u|<h}$ with $h > 0$,
- Gaussian kernel: $K(u) = \frac{1}{\sqrt{2\pi}}\exp(-\frac{u^2}{2})$.

Remark 1.4.1 At first, a "visual test" can be conducted, where the estimated density is plotted and compared to the density of a normal distribution with the same mean and variance as the sample. Then some mathematical quantities can be defined to measure the discrepancies between the curves, and then some statistical tests can be defined to accept or reject the normality assumption at a certain confidence level. Figure 1.3 is a histogram for the density estimated with a rectangular kernel and compares these values to the ones obtained for a normal distribution with the same mean and variance as the observations.

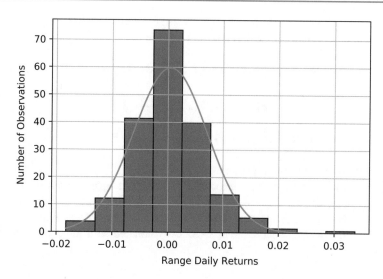

Fig. 1.3 Histogram daily returns, DAX 30 Index

1.4.2   Tests Based on Cumulative Distribution Function Estimates

Let X and $(X_i)_{i \in \mathbb{N}}$ be i.i.d. random variables with the same laws. Let $F_n(x) = \frac{1}{n} \sum_{i=1}^{i=n} 1_{X_i \leq x}$ and $\|F_n(x) - F(x)\|_\infty = \sup_x | F_n(x) - F(x) |$.

Theorem 1.4.2

$$\forall x \in \mathbb{R}, F_n(x) \to F(x)$$

and

$$\forall x \in \mathbb{R}, \sqrt{n}\big(F_n(x) - F(x)\big) \xrightarrow{Law} \mathcal{N}\big(0, F(x)[1 - F(x)]\big).$$

Proof We consider the random variables $Z_i = 1_{X_i \leq x}$.

As $\mathbf{E}(Z_i) = F(x)$ the first result follows from the law of large numbers and as $\mathbf{Var}(Z_i) = F(x)(1 - F(x))$ the second result is a consequence of the central limit theorem.   □

Theorem 1.4.3 (Glivenko–Cantelli) *Under certain regularity conditions,*

$$\|F_n(x) - F(x)\|_\infty \to 0 \, a.s.$$

For a proof see Dacunha-Castelle and Duflo [32].

Remark 1.4.2 This uniform convergence result is stronger than the simple convergence result mentioned in the first part of Theorem 1.4.2.

Theorem 1.4.4 (Kolmogorov–Smirnov) *Under certain regularity conditions,*

$$\sqrt{n}\|F_n(x) - F(x)\|_\infty \xrightarrow{Law} K,$$

where K is Kolmogorov's law (which is independent of F).

For a proof, see Dacunha-Castelle and Duflo [33].

Several goodness of fit tests are based on the Kolmogorov–Smirnov theorem, where F is chosen as a normal cumulative distribution function with the same mean and variance as the sample observed. Such a comparison is done in Fig. 1.4. Several of these tests are available in SAS, R, Python and with some excel extended libraries:

- the Kolmogorov–Smirnov test,
- the Cramér–von Mises test,
- the Anderson–Darling test.

Remark 1.4.3 Pointwise convergence theorems such as Theorem 1.4.2 are not well suited to test the normality hypothesis, as it is not optimal to test the similarity between two curves just by comparing the values of two functions at a single point. The Kolmogorov–Smirnov result is much more appropriate for this matter.

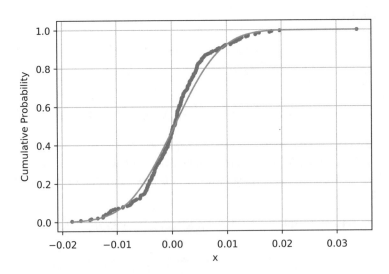

Fig. 1.4 Cumulative distribution function, DAX 30

1.4.3 Tests Based on Order Statistics

The last set of statistical tests is based on quantile statistics. As before, some "visual tests" can be conducted where the quantiles obtained from a sample are plotted against the quantiles from a normal distribution. Some statistical tests can then be conducted by measuring the distance between the points formed by these quantiles and a regression line.

Definition 1.4.1 (Quantile for a Distribution) If Z is a random variable, the α-**quantile** of Z is defined by $q_Z(\alpha) = \inf_x \{x, P(Z \leq x) \geq \alpha\}$.

Proposition 1.4.1 *If Z is a random variable, then $P(Z \leq q_Z(\alpha)) \geq \alpha$.*

Proof Let $\alpha > 0$ and $A = \{x, P(Z \leq x) \geq \alpha\}$, then $q_Z(\alpha) = \inf_{x \in A} x$ and

$$P(Z \leq q_Z(\alpha)) = P(Z \in] - \infty, \inf_{x \in A} x]).$$

Now, by definition of the inf we have $] - \infty, \inf_{x \in A} x] = \bigcap_{x \in A}] - \infty, x]$. So,

$$P(Z \in] - \infty, \inf_{x \in A} x]) = P(\bigcap_{x \in A} \{Z \in] - \infty, x]\}).$$

As P is a probability, we also have

$$P(\bigcap_{x \in A} \{Z \in] - \infty, x]\}) = \liminf_{x \in A} P(Z \in] - \infty, x])$$

but for each $x \in A$, $P(Z \in] - \infty, x]) \geq \alpha$. So, $\liminf_{x \in A} P(Z \in] - \infty, x]) \geq \alpha$ and thus $P(Z \leq q_Z(\alpha)) \geq \alpha$. □

Remark 1.4.4 If $Z \sim \mathcal{N}(0, 1)$ then $q_Z(0.5) = 0$ and $q_Z(97.5\%) \approx 1.96$.

Definition 1.4.2 (Quantile for a Sample) If $z = (z_1, z_2, \cdots, z_n)$ is a sample, the **quantile for the sample** z is defined as the quantile for the empirical distribution $\frac{1}{n} \sum_{i=1}^n \delta_{z_i}(\cdot)$.

Remark 1.4.5 If there are ten observations $\{z_i = i\}_{i \in [\![1,10]\!]}$ then $\hat{q}_z(1) = 10$ and $\forall i \in \{1, \cdots, 9\} \ \forall \alpha \in]\frac{i}{10}, \frac{i+1}{10}] \ \hat{q}_z(\alpha) = i + 1$.

Exercise 1.4.1 Show that if F_Z is invertible, then $q_Z(\alpha) = F_Z^{-1}(\alpha)$.

Exercise 1.4.2 Show that if F_Z is invertible, then $F_Z(Z) \sim U([0, 1])$.

Proposition 1.4.2 *If $Z_1 \sim \mathcal{N}(m_1, \sigma_1^2)$ and $Z_2 \sim \mathcal{N}(m_2, \sigma_2^2)$ then*

$$\{(q_{Z_1}(\alpha), q_{Z_2}(\alpha)), \alpha \in [0, 1]\}$$

is a line.

Proof Let $x = q_{Z_1}(\alpha)$, $y = q_{Z_2}(\alpha)$ and $Z \sim \mathcal{N}(0, 1)$. Then
$x = q_{Z_1}(\alpha) \implies P(Z_1 \leq x) = \alpha \implies P(m_1 + \sigma_1 Z \leq x) = \alpha \implies$
$P(Z \leq \frac{x-m_1}{\sigma_1}) = \alpha \implies P(m_2 + \sigma_2 Z \leq m_2 + \sigma_2 \frac{x-m_1}{\sigma_1}) = \alpha \implies P(Z_2 \leq$
$m_2 + \sigma_2 \frac{x-m_1}{\sigma_1}) = \alpha$
$\implies q_{Z_2}(\alpha) = m_2 + \sigma_2 \frac{x-m_1}{\sigma_1}$ so, $y = m_2 + \sigma_2 \frac{x-m_1}{\sigma_1}$.   □

Theorem 1.4.5 (Order Statistics and QQ-Plot Test) *Let $Z \sim \mathcal{N}(0, 1)$ and $z(n) = (z_1, z_2, \cdots, z_n)$ be a sample for the r.v. Z. Then when n is large, $\hat{q}_{z(n)}(\frac{i}{n}) \approx q_Z(\frac{i}{n})$ and if Z is a normal law, the points $(q_Z(\frac{i}{n}), \hat{q}_{z(n)}(\frac{i}{n}))$ should be "concentrated" around a line (called Henry's line).*

For a proof, see Hunter [52].

Remark 1.4.6 The test based on the distance to Henry's line is called the Shapiro–Wilk test.

Example 1.4.2 (QQ-Plot for the DAX 30 daily returns in 2017) When plotting in Fig. 1.5 the 251 quantiles for the DAX 30 daily returns against the quantiles of a normal distribution, the resulting points seem to lay close to a line. The fact that the points get further from the line at the extremities is linked to the fact that the empirical distribution has tails which are fatter than expected under the normality assumption.

1.4.4 Parameter Estimation and Confidence Intervals

For Gaussian distributions, some estimators can be built for the mean and the standard deviation. Some confidence intervals can also be built for the mean when the standard deviation is unknown. We present here these estimators, and recall some properties of the **chi-squared** and **Student distributions**.

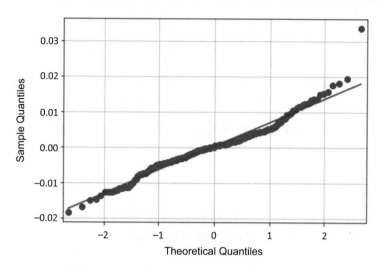

Fig. 1.5 QQ Plot, DAX 30

Definition 1.4.3 (Chi-Squared Distribution)

If $X \sim \mathcal{N}(0, 1)$ then the law of X^2 is called a **chi-squared distribution with one degree of freedom** and is denoted $\chi^2(1)$. If $X_i \sim \mathcal{N}(0, 1)$ i.i.d. then the law of $\sum_{i=1}^{i=n} X_i^2$ is called a **chi-squared distribution with n degrees of freedom** and is denoted $\chi^2(n)$.

Theorem and Definition 1.4.1 (Non-central Chi-Squared Distributions)

If $X_1 \sim \mathcal{N}(M_1, \mathrm{Id}_{\mathbb{R}^d})$ and $X_2 \sim \mathcal{N}(M_2, \mathrm{Id}_{\mathbb{R}^d})$, then $\|M_1\| = \|M_2\| \implies \|X_1\|^2$ and $\|X_2\|^2$ have the same law, which is written $\chi^2(d, \|M_1\|^2)$ and called a chi-squared distribution with d degrees of freedom and non-centrality parameter $\|M_1\|^2$.

Proof $\|M_1\| = \|M_2\| \implies \exists\, A$ orthonormal such that $AM_1 = M_2$. Let's consider AX_1, then X_1 has a Gaussian law $\implies AX_1$ has a Gaussian law, and $\mathbf{E}(AX_1) = A\mathbf{E}(X_1) = AM_1 = M_2$, and $\mathbf{Var}(AX_1) = \mathbf{Cov}(AX_1, AX_1) = A\mathbf{Cov}(X_1, X_1)A' = A\mathrm{Id}_{\mathbb{R}^d}A' = \mathrm{Id}_{\mathbb{R}^d}$. So, AX_1 and X_2 are both Gaussian with the same mean and variance. So, $AX_1 \sim X_2$ and consequently $\|AX_1\|^2 \sim \|X_2\|^2$. But $\|AX_1\|^2 = \|X_1\|^2$ (because A is orthonormal) so, we get $\|X_1\|^2 \sim \|X_2\|^2$. $\square$

Exercise 1.4.3 From the definition $\mathbf{Cov}(X, Y) = \mathbf{E}(XY') - \mathbf{E}(X)\mathbf{E}(Y)'$ show that

$$\mathbf{Cov}(AX, BY) = A\mathbf{Cov}(X, Y)B'$$

when A and B are matrices with the adequate dimensions. This formula will be used extensively when calculating the covariance of returns between two portfolios.

Definition 1.4.4 (Student's t-distribution) If $X \sim \mathcal{N}(0, 1)$ and $Z \sim \chi^2(n)$ are independent then we define

$$\frac{X}{\sqrt{\frac{Z}{n}}} \sim t(n).$$

$t(n)$ is called a **Student's t-distribution** with parameter n.

Remark 1.4.7 $t(n)$ is a symmetric distribution and it converges toward a $\mathcal{N}(0, 1)$ distribution as n increases. A Student's t-distribution can be defined through a density function for any parameter $\alpha > 0$ and thus the definition is not limited to an integer n. We mention the following results, usually derived from the density function:

- A Student's t-distribution presents fatter tails than a Gaussian distribution.
- If $n > 2$ the variance is defined and $\mathbf{Var}(t(n)) = \frac{n}{n-2}$.
- If $n > 3$ the skew is defined and $\mathbf{Skew}(t(n)) = 0$.
- If $n > 4$ the kurtosis is defined and $\mathbf{Kur}(t(n)) = 3 + \frac{6}{n-4}$.

Because of these properties, Student's t-distributions are sometimes considered as an alternative to Gaussian distributions, to model returns with fat tails.

Theorem 1.4.6 (Student's t-distribution for Confidence Intervals)
  Let $X = (X_1, X_2, \cdots, X_n)'$ with $X_i \sim \mathcal{N}(m, \sigma^2)$ i.i.d.
  Let $\overline{X} = \frac{1}{n}\sum_{i=1}^{i=n} X_i$ and $\hat{\sigma}(X) = \sqrt{\frac{1}{n}\sum_{i=1}^{i=n}(X_i - \overline{X})^2}$.
  Then,

(1) $\overline{X}$ *and* $\hat{\sigma}(X)$ *are independent,*
(2) $\sqrt{n}(\frac{\overline{X}-m}{\sigma}) \sim \mathcal{N}(0, 1)$,
(3) $n(\frac{\hat{\sigma}(X)}{\sigma})^2 \sim \chi^2(n-1)$, *which we can write as* $\frac{\|X-\overline{X}\|^2}{\sigma^2} \sim \chi^2(n-1)$,
(4) $\frac{\overline{X}-m}{\frac{\hat{\sigma}(X)}{\sqrt{n-1}}} \sim t(n-1)$, *which we can write as* $m = \overline{X} - \frac{\hat{\sigma}(X)}{\sqrt{n-1}}t(n-1)$.

Proof (Hint) We just need to show the result for $X_i \sim \mathcal{N}(0, 1)$ i.i.d.

Let $h : \mathbb{R}^n \longrightarrow \mathbb{R}^{n+1}$ be defined by $h : \begin{pmatrix} X_1 \\ \vdots \\ X_n \end{pmatrix} \longrightarrow \begin{pmatrix} \overline{X} \\ X - \overline{X}1_n \end{pmatrix}$ where 1_n

is the vector of $\mathbb{R}^n$ with all components equal to 1. Then, X Gaussian and h linear $\Rightarrow h(X)$ is Gaussian. Also, $\forall i$, $\mathbf{Cov}(\overline{X}, X_i - \overline{X}) = 0 \Rightarrow \overline{X}$ and $X - \overline{X}1_n$ are independent because for Gaussian vectors zero covariance implies independence. Note that, when calculating the scalar product $\langle , \rangle$ in $\mathbb{R}^n \langle X - \overline{X}1_n, 1_n \rangle = 0 \Rightarrow X - \overline{X}1_n \in (\mathbb{R}1_n)^{\perp}$. Now, $\forall u \in (\mathbb{R}1_n)^{\perp}$,

$$\mathbf{Var}(\langle u, X - \overline{X}1_n \rangle) = \mathbf{Var}(\langle u, X \rangle - \overline{X}\langle u, 1_n, \rangle)$$

$$= \mathbf{Var}(\langle u, X \rangle) = u'\mathbf{Var}(X)u = \|u\|^2.$$

So, $X - \overline{X}1_n$ is in a subspace E_{n-1} of dimension $n-1$ of $\mathbb{R}^n$ with law $\mathcal{N}(0, \mathrm{Id}_{E_{n-1}})$, which implies that $\|X - \overline{X}1_n\|_n^2 \sim \chi^2(n-1)$. □

Remark 1.4.8 A Gaussian variable can be defined in an intrinsic way, as follows.

If $(V, \langle \cdot, \cdot \rangle)$ is a Euclidean space, then X is a Gaussian vector of expectation $\vec{M} \in V$ and variance-covariance the quadratic form $\Phi_X(\cdot, \cdot)$ of V iff

- X takes its values in V almost P-surely,
- $\mathbf{E}(X) = \vec{M}$,
- $\forall u \in V$, $\langle u, X \rangle$ is a normal law of variance $\Phi_X(u, u)$,

and then we write $X \sim \mathcal{N}(\vec{M}, \Phi_X(\cdot, \cdot))$.

Example 1.4.3 (DAX 30 Returns for 2017) When applying the results above to build some estimators for the expectation and the standard deviation of the daily returns of the DAX 30, we get $\overline{X} = 0.045\%$ and $\frac{\hat{\sigma}(X)}{\sqrt{n-1}} = 0.042\%$. So, at confidence level 95% the expected daily rate of return $m = \mathbf{E}(X)$ is in the interval $[-0.03\%, 0.13\%]$.

Remark 1.4.9 (Student's t-distribution and Normal Distribution) From $\frac{\sqrt{n-1}(\overline{X}-m)}{\hat{\sigma}(X)}$ $\sim t(n-1)$ and the fact that $\hat{\sigma}(X)$ converges towards σ as n increases, we can deduce that when n is large a Student distribution is very similar to a normal distribution $\mathcal{N}(0, 1)$.

The definition of a Student's t-distribution can be extended to multidimensional space, as the ratio of a Gaussian vector and an independent chi-square distribution, as explained below. More details on the topic can be found in Kotz and Nadarajah [58].

Definition 1.4.5 (Multivariate t-distribution) Let Z and X be two independent random variables with $Z \sim \chi^2(n)$ and $X \sim \mathcal{N}(0, \Sigma_d)$, with Σ_d invertible. Then if we define

$$\frac{X}{\sqrt{\frac{Z}{n}}} \sim t(n, \Sigma_d).$$

$t(n, \Sigma_d)$ is called the **multivariate t-distribution** with parameters n and Σ_d.

1.5   Market Data with Python

1.5.1   Data Extraction for the DAX 30

The Python code used to extract historical prices from yahoo.finance for the DAX 30 Index, starting from 1 January 2000, and to print these prices for the five most recent dates is given in Listing 1.1:

Listing 1.1 Python. Data extraction for the DAX 30 Index

```
1 import numpy as np
2 import pandas as pd
3 import pandas_datareader.data as web
4 DAX = web.DataReader(name='^GDAXI', data_source='yahoo', start='
      2000-1-1')
5 DAX.tail() # visualisation of the 5 most recent data
6 # Other useful commands: DAX.head(), DAX.tail(), DAX.info()
```

The prices extracted for the five most recent dates are printed in Table 1.1.

1.5.2   Statistical Analysis for the DAX 30

The Python code to analyse the DAX 30 Index daily closing prices for the year 2017 and to produce the various Python figures of this chapter as well as the statistics of Skew and Kurtosis and the comparisons to a normal distribution is given in Listing 1.2:

Table 1.1 DAX 30: Five most recent prices

Date	High	Low	Open	Close	Volume	Adj Close
2018-11-30	11315.30	11208.60	11311.66	11257.24	109013100.0	11257.24
2018-12-03	11566.97	11457.61	11534.75	11465.46	101248500.0	11465.46
2018-12-04	11442.19	11330.44	11429.82	11335.32	83807700.0	11335.32
2018-12-05	11266.28	11177.15	11204.32	11200.24	73386400.0	11200.24
2018-12-06	11063.44	10884.48	11053.58	10908.13	0.0	10908.13

Listing 1.2 Python. Statistical analysis for the DAX 30

```python
# 1) Graph for the DAX 30 closing prices
import numpy as np
import pandas as pd
import pandas_datareader.data as web
import matplotlib.pyplot as plt
DAX = web.DataReader(name='^GDAXI', data_source='yahoo', start='
    2017-1-1', end='2017-12-31')
plt.plot(DAX['Close'])
plt.legend(['DAX'], loc=2)
plt.grid(True)
plt.xlabel('Dates')
plt.ylabel('Closing Prices')
plt.show()

# 2) Graph for the DAX daily returns
DAX['Return'] = DAX['Close']/DAX['Close'].shift(1)-1
plt.plot(DAX['Return'], 'b', lw=0.5) # thickness of the line
plt.plot(DAX['Return'], 'r.') # the points appear in red
plt.grid(True)
plt.xlabel('Dates')
plt.ylabel('Daily Returns')
plt.show()

# 3) Table of descriptives for the returns
from scipy import stats
from scipy.stats import norm
print ("Stat Returns: ", stats.describe(DAX['Return'][1:]))
# 4) Histogram of the daily returns with fitting to a normal
    curve
import matplotlib.mlab as mlab
Return = DAX['Return'][1:] # disregard the first missing value
plt.hist(Return, edgecolor='black', density=True, bins=10)
# parameters of the normal curve
meanR = float(np.mean(Return))
sdevR = float(np.std(Return, ddof=1))
minR = float(np.min(Return))
maxR = float(np.max(Return))
x = np.linspace(minR, maxR, 100)
plt.plot(x, norm.pdf(x, meanR, sdevR))
plt.grid(True)
plt.xlabel('Range Daily Returns')
plt.ylabel('Number of Observations')
plt.show()

# 5) Comparison Cumulative Distribution Functions
x = np.sort(DAX['Return'][1:]) # disregard the first missing
    value and sort the data
y = np.arange(1, len(x)+1) / float(len(x))
meanR = np.mean(x)
sdevR = np.std(x)
y1 = norm.cdf((x-meanR)/sdevR)
plt.plot(x, y, marker='.', linestyle='none')
```

```
50  plt.plot(x,y1)
51  plt.grid(True)
52  plt.xlabel('x')
53  plt.ylabel('Cumulative Probability')
54  plt.show()
55
56  # 6) Henry's line, QQ Plot
57  import statsmodels.api as sm
58  data = DAX['Return'][1:] # disregard the first missing value
59  sm.qqplot(data, line='s')
60  plt.grid(True)
61  # Other useful commands: DAX['Return'].dropna() to eliminate the
        missing values
```

A Few References

1. Bera, A., & Jarque, C. (1987). A test for normality of observations and regression residuals. *International Statistical Review, 55*, 163–172.
2. Dacunha-Castelle, D., & Duflo, M. (1986). *Probability and statistics* (Vol. I). New York: Springer.
3. Dacunha-Castelle, D., & Duflo, M. (1986). *Probability and statistics* (Vol. II). New York: Springer.
4. Darren D., Droettboom M., Firing E., Hunter J., & the Matplotlib Development Team. (2012). https://matplotlib.org/gallery/index.html.
5. Hilpisch, Y. (2014). *Python for finance*. Cambridge: O'Reilly.
6. Hunter, D. R. (2006). *Statistics 553 asymptotic tools*. Lecture Notes Fall 2006 Chapter 6.
7. Kotz, M., & Naradajah, S. (2004). *Multivariate t-distributions and their applications*. Cambridge: Cambridge University Press.
8. Parzen, E. (1962). On estimation of a probability density function and mode. *Financial Analysts Journal. Annals of Mathematical Statistics, 33*, 1065–1076.
9. Steiner, B. (2012). *Key financial market concepts* (2nd ed.). New York: Financial Times/Prentice Hall Ltd.

Utility Functions and the Theory of Choice

2

When payouts are deterministic, investor preferences are easy to determine, and if the payout of asset A is twice the payout of asset B, then the price of A is twice the price of B. Now, when payouts are random, determining the criteria of choice between two investments is more complex, and if the expected payout of asset A is twice the expected payout of asset B, the price of A is not necessarily twice the price of B. Furthermore, when choosing between two assets, beyond the mathematical expectations of the payouts, the variances and the whole distribution of the payouts are usually considered. For a random payout X, this analysis leads us to look not only at its expectation $\mathbf{E}(X)$ but at something of the form $\mathbf{E}\big(u(X)\big)$, where u is called a **utility function**. In this context, $\mathbf{E}\big(u(X)\big)$ is the objective to maximise when choosing between different random payouts X. The function u is chosen to reflect the preferences of the investors and it will be shown in this chapter that the appetite or aversion to risk is linked to the convexity or concavity of the function u. An extensive literature exists in economics about utility functions and their applications. John von Neumann and Oskar Morgenstern proved that, from a mathematical point of view, individuals whose preferences satisfy four particular axioms have a utility function. Gerard Debreu (Nobel price in Economics 1983) and Kenneth Arrow (Nobel price in Economics 1972) produced additional landmark results, defining the equilibrium in an economy where agents act upon some utility functions. This chapter is only a very brief introduction to the topic, and is included to make the link with the mean-variance criteria which is used in the rest of the book.

2.1 Utility Functions and Preferred Investments

The mathematical definition of a utility function is given here, and the way an investment choice is made according to such a utility function is described.

© Springer Nature Switzerland AG 2020
P. Brugière, *Quantitative Portfolio Management*, Springer Texts in Business and Economics, https://doi.org/10.1007/978-3-030-37740-3_2

Definition 2.1.1 (Utility Functions) A **utility function** is defined as a function $u :$ $\mathbb{R} \longrightarrow \mathbb{R}$ such that u is continuous, strictly increasing and twice differentiable.

Remark 2.1.1 Based on its definition, any utility function u is invertible.

Definition 2.1.2 (Preferred Investment, Certain Equivalent and Risk Premium) If X and Y are two random payoffs:

- $X \succ Y$ (X is **preferred** to Y for u) $\Leftrightarrow \mathbf{E}(u(X)) \geq \mathbf{E}\big(u(Y)\big)$.
- $X \equiv Y$ (X is **equivalent** to Y for u) $\Leftrightarrow \mathbf{E}(u(X)) = \mathbf{E}\big(u(Y)\big)$.

Let $C_u(X)$ be the constant defined by $u\big(C_u(X)\big) = \mathbf{E}\big(u(X)\big)$. Then

- $C_u(X)$ is called the **certain equivalent** to X,
- $\Pi_u(X) = \mathbf{E}(X) - C_u(X)$ is called the **risk premium** for X.

From now on, a risk taker is defined as someone who prefers, for any random payout X, to receive X rather than $\mathbf{E}(X)$, a risk adverse person as someone who prefers to receive $\mathbf{E}(X)$ rather than X and a risk neutral person as someone who is indifferent to receiving X or $\mathbf{E}(X)$. This translates in terms of utility functions in the following way:

Definition 2.1.3

- for a **risk taker** $X \succ \mathbf{E}(X)$, i.e. $\mathbf{E}(u_r(X)) \geq u_r(\mathbf{E}(X))$,
- for a **risk adverse person** $\mathbf{E}(X) \succ X$, i.e. $\mathbf{E}(u_a(X)) \leq u_a(\mathbf{E}(X))$,
- for a **risk neutral person** $X \equiv \mathbf{E}(X)$, i.e. $\mathbf{E}(u_n(X)) = u_n(\mathbf{E}(X))$.

According to this definition, the risk premium is positive when an investor is risk adverse. In this case, the investor expects to be remunerated for the risk he takes by purchasing the asset at a discount to its expected payout. Figure 2.1 illustrates the concept.

2.1.1   Risk Appetite and Concavity

We explain in this section how the risk premium is related mathematically to the concavity of the utility function that the investor uses to modelise his preferences.

Exercise 2.1.1 We leave to the reader the demonstrations of the following properties:

u convex $\Leftrightarrow \forall X, u(\mathbf{E}(X)) \leq \mathbf{E}(u(X)) \Leftrightarrow \forall X, \Pi_u(X) \leq 0 \Leftrightarrow$ risk taker,
u concave $\Leftrightarrow \forall X, u(\mathbf{E}(X)) \geq \mathbf{E}(u(X)) \Leftrightarrow \forall X, \Pi_u(X) \geq 0 \Leftrightarrow$ risk adverse,
u affine $\Leftrightarrow \forall X, u(\mathbf{E}(X)) = \mathbf{E}(u(X)) \Leftrightarrow \forall X, \Pi_u(X) = 0 \Leftrightarrow$ risk neutral.

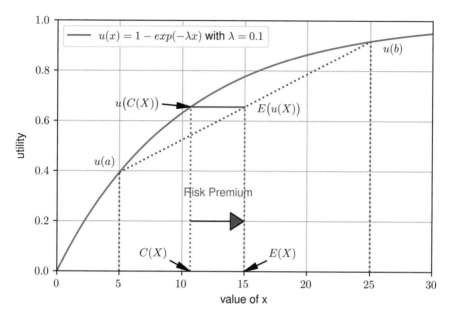

Fig. 2.1 Equivalent certain and risk premium for a concave utility function

The following theorem links the notion of risk premium to a measure of concavity of the utility function, given by a formula based on its first and second-order derivatives.

Theorem 2.1.1 (Risk Aversion Measure) *Let u and v be two utility functions then:*

$$(\forall X \text{ discrete r.v., } \Pi_u(X) \geq \Pi_v(X)) \Leftrightarrow \left(\forall a \in \mathbb{R}, \; -\frac{u''(a)}{u'(a)} \geq -\frac{v''(a)}{v'(a)}\right).$$

Remark 2.1.2 (Concavity and Risk Aversion) The derivatives are strictly positive (and thus do not cancel) as the functions are strictly increasing and $-\frac{u''}{u'}$ defines the risk aversion measure for u. $\forall \lambda \neq 0$, λu and u have the same risk aversion measure. The risk aversion measure defined here differs from the geometric definition of curvature which is given by $\dfrac{u''}{(1+u'^2)^{\frac{3}{2}}}$.

Proof (Hint) Let u be a twice differentiable utility function. Let's demonstrate the first implication $\Rightarrow$ of Theorem 2.1.1. Let X_h^a be the random payout defined by $P(X_h^a = a) = \frac{1}{2}$ and $P(X_h^a = a + h) = \frac{1}{2}$. Let $C_u^a(h)$ be the certain equivalent for this random payout. Then,

$$u(C_u^a(h)) = \mathbf{E}(u(X_h^a)) = \frac{1}{2}u(a) + \frac{1}{2}u(a + h). \tag{2.1.1}$$

By taking the derivatives of Eq. (2.1.1), we get

$$u'(C_u^a(h))C_u^{a'}(h) = \frac{1}{2}u'(a+h) \tag{2.1.2}$$

and

$$u''(C_u^a(h))(C_u^{a'}(h))^2 + u'(C_u^a(h))C_u^{a''}(h) = \frac{1}{2}u''(a+h) \tag{2.1.3}$$

and by taking $h = 0$ we get,

(1) $C_u^a(0) = a$ from Eq. (2.1.1),
(2) $C_u^{a'}(0) = \frac{1}{2}$ from Eq. (2.1.2),
(3) $C_u^{a''}(0) = \frac{1}{4}\frac{u''(a)}{u'(a)}$ from Eq. (2.1.3).

Now,

$$\Pi_u \geq \Pi_v \Rightarrow \forall a, \forall h, \Pi_u(X_h^a) \geq \Pi_v(X_h^a)$$
$$\Rightarrow \forall h, \mathbf{E}(X_h^a) - C_u^a(h) \geq \mathbf{E}(X_h^a) - C_v^a(h)$$
$$\Rightarrow -C_u^{a''}(0) \geq -C_v^{a''}(0) \text{ (as at 0 the value of the two functions and their}$$
$$\text{first derivatives are equal)}$$
$$\Rightarrow -\frac{u''(a)}{u'(a)} \geq -\frac{v''(a)}{v'(a)} \text{ (and this is true for all } a\text{).}$$

Let's demonstrate the second implication $\Leftarrow$ of Theorem 2.1.1.
Let X be a discrete variable with $P(X = a_i) = p_i$.

$$\Pi_u(X) \geq \Pi_v(X) \Leftrightarrow C_v(X) - C_u(X) \geq 0$$
$$\Leftrightarrow v^{-1}(\mathbf{E}(v(X))) - u^{-1}(\mathbf{E}(u(X))) \geq 0$$
$$\Leftrightarrow v^{-1}\left(\sum_{i=1}^{i=n} p_i v(a_i)\right) - u^{-1}\left(\sum_{i=1}^{i=n} p_i u(a_i)\right) \geq 0$$
$$\Leftrightarrow \sum_{i=1}^{i=n} p_i v(a_i) \geq v \circ u^{-1}\left(\sum_{i=1}^{i=n} p_i u(a_i)\right) \text{ (as } v \text{ is increasing),}$$

which is true if $v \circ u^{-1}$ is convex, according to Jensens's inequality.

Let's calculate $(v \circ u^{-1})''$ to prove that $v \circ u^{-1}$ is convex.

$$(v \circ u^{-1})' = (v' \circ u^{-1})(u^{-1})',$$

$$(v \circ u^{-1})'' = (v'' \circ u^{-1})[(u^{-1})']^2 + (v' \circ u^{-1})(u^{-1})''.$$

Now, $(u^{-1})' = \frac{1}{u' \circ u^{-1}}$ and $\left(\frac{1}{u' \circ u^{-1}}\right)' = -\frac{(u'' \circ u^{-1})(u^{-1})'}{(u' \circ u^{-1})^2} = -\frac{(u'' \circ u^{-1})}{(u' \circ u^{-1})^3}.$

So,

$$(v \circ u^{-1})'' \geq 0$$

$$\Leftrightarrow \frac{(v'' \circ u^{-1})}{(u' \circ u^{-1})^2} - (v' \circ u^{-1})\frac{(u'' \circ u^{-1})}{(u' \circ u^{-1})^3} \geq 0$$

$$\Leftrightarrow (v'' \circ u^{-1}) \geq (v' \circ u^{-1})\frac{(u'' \circ u^{-1})}{(u' \circ u^{-1})}$$

$$\Leftrightarrow -\frac{(v'' \circ u^{-1})}{(v' \circ u^{-1})} \leq -\frac{(u'' \circ u^{-1})}{(u' \circ u^{-1})},$$

which holds as we assumed that $-\frac{v''}{v'} \leq -\frac{u''}{u'}$. $\qquad\square$

Remark 2.1.3 To model risk aversion, a concave utility function u is chosen. A frequent choice is $u(a) = 1 - \exp(-\lambda a)$. In this case, $-\frac{u''}{u'} = \lambda$ and the parameter λ determines the risk aversion of the investor.

2.2 Gaussian Laws and Mean-Variance Implications

With the utility function u and for an investment of W_0, the strategy π selected is the one which maximises $\mathbf{E}(u(W_T(\pi)))$, where $W_T(\pi) = W_0(1 + R_T(\pi))$ and $R_T(\pi)$ is the return generated by the strategy π.

If we assume for the return of the strategy that $R_T(\pi) \sim N(m_\pi, \sigma_\pi^2)$, then

$$\mathbf{E}(u(W_T(\pi))) = \mathbf{E}(u(W_0 + W_0 R_T(\pi))) = \mathbf{E}(u(W_0 + m_\pi W_0 + \sigma_\pi W_0 Z)),$$

with $Z \sim N(0, 1)$.

If we define $U(m, \sigma) = \mathbf{E}(u(W_0 + m W_0 + \sigma W_0 Z))$, then

$$\max_\pi \mathbf{E}(u(W_T(\pi))) = \max_\pi U(m_\pi, \sigma_\pi).$$

So in practice the model consists in maximising $\max_\pi U(m_\pi, \sigma_\pi)$, where U is derived from a utility function.

Remark 2.2.1 For the DAX, if we assume that we are at an equilibrium, where investors are therefore indifferent (when applying their utility functions to the expected returns) between investing into a 1 year zero coupon bond, which yields approx 0%, or into the DAX, which has a yearly expected return of approx $0.04\% \times 251 = 10\%$ (if we consider that past performance is a good indicator of future performance), then the Risk Premium of the DAX comes out as 10%. Basically, risk adverse investors require an excess expected return of 10% to invest into a risky asset, rather than into a 1 year risk-free bond.

From now on, we will modelise the preferences directly with a function U, of m and σ, and we will assume that U is increasing in m and decreasing in σ, which means that

- if an investor has the choice between two strategies with the same expected returns, he prefers the one with the smallest variance of the returns,
- if an investor has the choice between two strategies with the same variance of the returns, he prefers the one with the highest expected return.

In practice, an investor defines the level of risks he accepts and based on this finds the efficient portfolio (which maximises the expected return).

Based on this, we can give the definition of an efficient investment strategy.

2.3   Efficient Investment Strategies

Let $R(\pi)$ be the random return for the investment strategy π.

Definition 2.3.1 (Efficient Investment Strategy) An investment strategy π_e is **efficient** iff for all investment strategies π,

$$\begin{cases} \mathbf{E}(R(\pi)) > \mathbf{E}(R(\pi_e)) \Rightarrow \sigma(R(\pi)) > \sigma(R(\pi_e)) \text{ and} \\ \mathbf{E}(R(\pi)) = \mathbf{E}(R(\pi_e)) \Rightarrow \sigma(R(\pi)) \geq \sigma(R(\pi_e)). \end{cases}$$

Proposition 2.3.1 *An efficient investment strategy π_e is a solution of*

$$\begin{cases} \inf_{\pi} \sigma(R(\pi)) \\ \mathbf{E}(R(\pi)) = \mathbf{E}(R(\pi_e)). \end{cases}$$

Proof (Hint) $\sigma(R(\pi))$ is a continuous function of π which tends to infinity as $\|\pi\|$ tends to infinity, so the inf is reached as the inf is to be found in a compact set of $\mathbb{R}^d$. If the inf was less than $\sigma(R(\pi_e))$ then we could find π which reaches the inf such that

$$\sigma(R(\pi)) < \sigma(R(\pi_e)) \text{ and } \mathbf{E}(R(\pi)) = \mathbf{E}(R(\pi_e)),$$

which would contradict the fact that π_e is efficient.   □

Remark 2.3.1 (Notations and Geometric Interpretation) If we write

$$\mathcal{E} = \{\pi_e \text{ efficient}\} \text{ and } \mathcal{E}(\sigma, m) = \left\{ \begin{pmatrix} \sigma(R(\pi_e)) \\ \mathbf{E}(R(\pi_e)) \end{pmatrix}, \pi_e \in \mathcal{E} \right\},$$

then any point $\begin{pmatrix} \sigma(R(\pi)) \\ \mathbf{E}(R(\pi)) \end{pmatrix}$ is either on $\mathcal{E}(\sigma, m)$ or to the right of it.

A Few References

1. Arrow, K., & Debreu, G. (1954). Existence of an equilibrium for a competitive economy. *Econometrica, 22*(3), 265–290.
2. Debreu, G. (1983). *Mathematical economics.* Cambridge: Cambridge University Press.
3. Von Neumann, J., & Morgenstern, O. (1947). *Theory of games and economic behavior.* Princeton: Princeton University Press.

The Markowitz Framework

3

We present here the mathematical framework under which the "Markowitz problem" of maximising the expected return of a portfolio under a risk constraint is solved.

3.1 Investment and Self-Financing Portfolios

We consider an economy with d risky assets, whose returns are modelled by a Gaussian law of mean vector M and variance-covariance matrix Σ. A portfolio is represented by a vector π, whose components correspond to the relative value of each asset, compared to an initial wealth x_0. We demonstrate how to calculate the expectations, standard deviations and correlations of the returns of any portfolios, by simple algebraic calculations on the allocation vectors π. Two types of portfolios are considered: **self-financing portfolios** and **investment portfolios**. For self-financing portfolios, at inception, the total cost of the long positions is financed by the total cash generated by the short positions, therefore no initial capital is required to put in place these long/short strategies. For investment portfolios, even if both long and short positions can be present, there is an initial cost x_0 to create the portfolio and the vector of allocation corresponds to the relative value of each asset position compared to x_0.

© Springer Nature Switzerland AG 2020
P. Brugière, *Quantitative Portfolio Management*, Springer Texts in Business
and Economics, https://doi.org/10.1007/978-3-030-37740-3_3

3.1.1   Notations and Definitions

We denote by $(S_i)_{i \in [\![1,d]\!]}$ the d risky assets of the economy and use the following notations.

Notations

- S_t^i, the value of asset i at time t. Here we consider only two instants (0 and T),
- q_i, the number of shares i held ($q_i > 0$) or shorted ($q_i < 0$) at time 0,
- R_T^i, the return of asset i between 0 and T, i.e. $R_T^i = \frac{S_T^i}{S_0^i} - 1$,

- $R_T = \begin{pmatrix} R_T^1 \\ \vdots \\ R_T^d \end{pmatrix}$, the vector of returns of the d assets between 0 and T,

- $M = \mathbf{E}(R_T)$, the vector of expected returns of the assets,
- $\Sigma = \mathbf{Cov}(R_T, R_T)$, the variance-covariance matrix for the returns of the assets.

Remark 3.1.1 In the model, the possibility of having $q_i < 0$ is assumed and such a position in asset i is called a **short position**. In practice, this short position can be created by borrowing at inception the asset i from a market participant (a stock lender) and selling it immediately in the market at price S_0^i. At time T, the asset is repurchased in the market at price S_T^i and immediately returned to the stock lender. This transaction, once completed, has not impacted the stock lender. For the fund manager, who borrowed and sold the stock before repurchasing it, the resulting economics is a net cash flow of $S_0^i - S_T^i$, which is the opposite of what is obtained from a "**long strategy**", consisting of purchasing the stock at time 0 and unwinding the position by a sale at time T.

3.1.2   Representations of the Portfolios

The notion of a self-financing portfolio is very useful when interpreting the many formulas that appear throughout this book. Also, for practitioners, self-financing portfolios can be viewed as derivative overlay strategies and the difference between the allocations of two distinct investment portfolios corresponds to the allocation of a self-financing portfolio.

Definition 3.1.1 (Investment Portfolio, Self-Financing Portfolio)

- An **investment portfolio** is a portfolio for which $\sum_{i=1}^{i=d} q_i S_0^i > 0$.

- A **self-financing portfolio** is a portfolio for which $\sum_{i=1}^{i=d} q_i S_0^i = 0$.

Property 3.1.1 (Representation of an Investment Portfolio)

(1) To each investment portfolio $(q_i, \cdots, q_d)$ we can associate the $(d+1)$-tuple $(x_0, \pi_1, \cdots, \pi_d)$ with $x_0 = \sum\limits_{i=1}^{i=d} q_i S_0^i$ and $\pi_i = \frac{q_i S_0^i}{x_0}$ and by doing so we get $x_0 > 0$ and $\sum\limits_{i=1}^{i=d} \pi_i = 1$.

(2) Conversely, to each $(d+1)$-tuple $(x_0, \pi_1, \cdots, \pi_d)$ for which $x_0 > 0$ and $\sum\limits_{i=1}^{i=d} \pi_i = 1$ we can associate, by inverse transformation, the d-tuple $(q_1, \cdots, q_d)$ with $q_i = \frac{\pi_i x_0}{S_0^i}$, which represents an investment portfolio of initial value $\sum\limits_{i=1}^{i=d} q_i S_0^i = x_0$.

Proof

(1) By definition of an investment portfolio $x_0 = \sum\limits_{i=1}^{i=d} q_i S_0^i > 0$, so $x_0 > 0$.

   Also, $\sum\limits_{i=1}^{i=d} \pi_i = \sum\limits_{i=1}^{i=d} \frac{q_i S_0^i}{x_0} = \frac{x_0}{x_0} = 1$.

(2) The transformation $\pi_i \longrightarrow q_i = \frac{\pi_i x_0}{S_0^i}$ is the inverse transformation of $q_i \longrightarrow \pi_i = \frac{q_i S_0^i}{x_0}$.

   The value of $\sum\limits_{i=1}^{i=d} q_i S_0^i$ is $\sum\limits_{i=1}^{i=d} \pi_i x_0 = \left(\sum\limits_{i=1}^{i=d} \pi_i \right) x_0 = x_0$. $\square$

Property 3.1.2 (Representation of a Self-Financing Portfolio) Let $x_0 > 0$ be a fixed number.

(1) To each self-financing portfolio $(q_i, \cdots, q_d)$ we can associate the d-tuple $(\pi_1, \cdots, \pi_d)$ with $\pi_i = \frac{q_i S_0^i}{x_0}$ and by doing so we get $\sum\limits_{i=1}^{i=d} \pi_i = 0$.

(2) Conversely, to each d-tuple $(\pi_1, \cdots, \pi_d)$ for which $\sum\limits_{i=1}^{i=d} \pi_i = 0$ we can associate by inverse transformation the d-tuple $(q_1, \cdots, q_d)$ with $q_i = \frac{\pi_i x_0}{S_0^i}$, which represents a self-financing portfolio.

Proof

(1) $\sum\limits_{i=1}^{i=d} \pi_i = \sum\limits_{i=1}^{i=d} \frac{q_i S_0^i}{x_0} = \frac{1}{x_0} \sum\limits_{i=1}^{i=d} q_i S_0^i = 0$.

Table 3.1 Various portfolios

x_0	π_1	π_2	q_1	q_2	$W_0(x_0, \pi)$
100	0.5	-0.5	5	-10	0
1000	1	-1	100	-200	0
100	0.5	0.5	5	10	100
1000	0.5	0.5	50	100	1000

(2) The transformation $\pi_i \longrightarrow q_i = \frac{\pi_i x_0}{S_0^i}$ is the inverse transformation of $q_i \longrightarrow$
$\pi_i = \frac{q_i S_0^i}{x_0}$.

The value of $\sum_{i=1}^{i=d} q_i S_0^i$ is $\sum_{i=1}^{i=d} \pi_i x_0 = \left(\sum_{i=1}^{i=d} \pi_i \right) x_0 = 0$.

$\square$

Notations

- Based on Property 3.1.1 we will now represent an investment portfolio by a $(d+1)$-tuple $(x_0, \pi_1, \cdots, \pi_d)$ with $x_0 > 0$ and $\sum_{i=1}^{i=d} \pi_i = 1$.
- Based on Property 3.1.2 we will now represent a self-financing portfolio by a $(d+1)$-tuple $(x_0, \pi_1, \cdots, \pi_d)$ with $x_0 > 0$ and $\sum_{i=1}^{i=d} \pi_i = 0$.
- The value at time 0 of the portfolio $(x_0, \pi_1, \cdots, \pi_d)$ is denoted $W_0(x_0, \pi)$.
- The value at time T of the portfolio $(x_0, \pi_1, \cdots, \pi_d)$ is denoted $W_T(x_0, \pi)$.

Example 3.1.1 Table 3.1 represents various allocations (x_0, π) between two shares of initial values $S_0^1 = 10$ and $S_0^2 = 5$.

Definition 3.1.2 (Notional) For a self-financing portfolio $(x_0, \pi_1, \cdots, \pi_d)$, x_0 is called the **notional**.

3.1.3 Return of a Portfolio

Definition 3.1.3 We define the **return** of a portfolio (x_0, π) as

$$R_T(\pi) = \frac{W_T(x_0, \pi) - W_0(x_0, \pi)}{x_0}.$$

Proposition 3.1.1

- *For an investment portfolio this definition corresponds to the definition of the return of an asset.*
- *For a self-financing portfolio this quantity equals $\frac{W_T(x_0, \pi)}{x_0}$.*

Proof Trivial as for a self-financing portfolio we have $W_0(x_0, \pi) = 0$.   □

The following proposition establishes that the return of a portfolio is the average of the returns of its components and that the weights to be used in the averaging calculation are the weights of the stocks in the portfolio, i.e. the percentage of the wealth for an investment portfolio, and the percentage of the notional for a self-financing portfolio, which is invested in these stocks.

Proposition 3.1.2 (Return of a Portfolio) *For any investment or self-financing portfolio* (x_0, π) *the return over* $[0, T]$ *depends only on the strategy* π *and not on* x_0 *and satisfies*

$$R_T(\pi) = \sum_{i=1}^{i=d} \pi_i R_T^i.$$

Proof $W_T(x_0, \pi) = \sum_{i=1}^{i=d} x_0 \pi_i \dfrac{S_T^i}{S_0^i}.$

For an investment portfolio,

$$R_T(\pi) = \frac{W_T(x_0, \pi)}{x_0} - 1 = \sum_{i=1}^{i=d} \pi_i \frac{S_T^i}{S_0^i} - 1$$

$$= \sum_{i=1}^{i=d} \pi_i \frac{S_T^i}{S_0^i} - \sum_{i=1}^{i=d} \pi_i = \sum_{i=1}^{i=d} \pi_i \left(\frac{S_T^i}{S_0^i} - 1\right)$$

$$= \sum_{i=1}^{i=d} \pi_i R_T^i.$$

For a self-financing portfolio,

$$R_T(\pi) = \frac{W_T(x_0, \pi)}{x_0} = \sum_{i=1}^{i=d} \pi_i \frac{S_T^i}{S_0^i} - \sum_{i=1}^{i=d} \pi_i = \sum_{i=1}^{i=d} \pi_i \left(\frac{S_T^i}{S_0^i} - 1\right) = \sum_{i=1}^{i=d} \pi_i R_T^i,$$

which finishes the proof.   □

Proposition 3.1.3 *For any investment or self-financing portfolio* (x_0, π) *we have:*

(1) $R_T(\pi) = \pi' R_T,$
(2) $\mathbf{E}(R_T(\pi)) = \pi' \mathbf{E}(R_T),$
(3) when considering two portfolios π_P *and* π_Q *we have*

$$\mathbf{Cov}(R_T(\pi_P), R_T(\pi_Q)) = \pi_P' \mathbf{Cov}(R_T, R_T) \pi_Q.$$

Proof This results from the linearity of the expectation and the bilinearity of the covariance.   □

Notations From now on we write:

- $\sigma_\pi = \sigma\big(R_T(\pi)\big)$, the standard deviation of the return of the strategy π,
- $m_\pi = \mathbf{E}\big(R_T(\pi)\big)$, the expected return of the strategy π.

3.2   Absence of Arbitrage Opportunities

We define here the condition of **Absence of Arbitrage Opportunities (AAO)**, which, as will be demonstrated, implies the invertibility of the variance-covariance matrix Σ for the returns of the risky assets.

Definition 3.2.1 (AAO Conditions) The **AAO conditions** are satisfied iff the two properties below are satisfied:

- (AAO1) If two risk-free investment strategies exist, their rate of returns are the same.
- (AAO2) If a self-financing strategy is without risk, its return is zero.

Remark 3.2.1 (AAO1) means that it is not possible to have two distinct risk-free rates in the economy. (AAO2) means that it is not possible to create money for sure out of nothing. If these conditions were not satisfied, it would be possible to find strategies to become infinitely rich without risk, and no model would be required to devise investment strategies.

Remark 3.2.2 (AAO1) means that, if it is possible to build with the risky assets some risk-free investment portfolios, then these risk-free investment portfolios should all have the same return and if there is a risk-free asset of return r_0 in the economy, then the return for these risk-free investment portfolios should all be r_0.

Proposition 3.2.1 *If Σ is invertible then:*

*(1)   the **AAO** conditions are satisfied for the economy consisting of the risky assets,*
*(2)   if a risk-free asset of return r_0 is added to the economy, the **AAO** conditions are satisfied as well.*

Proof (1) The variance of a portfolio of risky asset is $\pi'\Sigma\pi$. So, if Σ is invertible it is not possible to build a risk-free portfolio with the risky assets and therefore the **AAO** conditions cannot be contradicted. (2) If a risk-free asset is added to the economy, the only risk-free rate is the rate of this asset and while (AAO1) is still valid (AAO2) is satisfied as well.   □

3.2.1   Analysis of the Variance-Covariance Matrix

As seen previously, having a positive definite matrix of variance-covariance Σ guarantees that the **AAO** conditions are satisfied. In practice, it will always be assumed that Σ is positive definite, but in this section we analyse what would happen if this was not the case.

Property 3.2.1 (Consequences If Σ Is Not Invertible) If Σ is not invertible then it is possible to build either an investment portfolio or a self-financing portfolio which is risk-free.

Proof If Σ is not invertible it is possible to find $\pi \neq 0$ such that $\pi'\Sigma\pi = 0$. Then, either $\pi'1_d = 0$, in which case π is a self-financing portfolio without risk, or $\pi'1_d \neq 0$, in which case $\frac{\pi}{\pi'1_d}$ is an investment portfolio without risk.   □

Remark 3.2.3 (Consequences If There Are Risk-Free Self-Financing Portfolios) If it is possible to build a risk-free self-financing portfolio then for the **AAO** conditions to be satisfied its return must be zero, which implies that this self-financing portfolio has a constant value of zero, and that therefore one of the risky asset can be replicated by the others.

Remark 3.2.4 (Consequences If There Are Risk-Free Investment Portfolios) If it is possible to build risk-free investment portfolios, then for the **AAO** conditions to be satisfied such risk-free investment portfolios must all have the same return. This return is then the risk-free rate, and any such portfolio can be considered as a risk-free asset. Then some risky assets can be replicated by this risk-free asset and other risky assets, and therefore are redundant in the economy consisting of the risky assets and of this risk-free rate.

As it does not make sense to introduce redundant assets to describe/generate the economy, the hypothesis will always be made in the model that Σ is invertible and in this case, as seen previously, the **AAO** conditions will be satisfied.

In the next chapters we will consider two successive models, with Σ invertible. In the first model, the economy has no risk-free asset. In the second model, there is an additional risk-free asset. The second economy is therefore the same as the initial economy but augmented by a risk-free asset.

3.2.2   The Correlation Matrix

Usually, people prefer to think in terms of a correlation matrix Λ and individual standard deviations $(\sigma_1, \cdots, \sigma_d)$ rather than in terms of a variance-covariance matrix Σ, whose values are more difficult to interpret. Of course, the link between the two is straightforward. Let diag(σ_i) be the diagonal matrix whose diagonal components are the σ_i, then the following properties are satisfied:

Proposition 3.2.2 *The correlation matrix Λ and the variance-covariance matrix Σ of the returns of the risky assets are linked by the relationship:*

$$\Sigma = \text{diag}(\sigma_i)\Lambda\text{diag}(\sigma_i). \tag{3.2.1}$$

Proof

$$\Lambda = \mathbf{Cov}\left(\begin{pmatrix} \frac{X_1}{\sigma_1} \\ \vdots \\ \frac{X_d}{\sigma_d} \end{pmatrix}, \begin{pmatrix} \frac{X_1}{\sigma_1} \\ \vdots \\ \frac{X_d}{\sigma_d} \end{pmatrix}\right) = \mathbf{Cov}\left(\text{diag}\left(\frac{1}{\sigma_i}\right)X, \text{diag}\left(\frac{1}{\sigma_i}\right)X\right),$$

so

$$\Lambda = \text{diag}\left(\frac{1}{\sigma_i}\right)\mathbf{Cov}(X, X)\text{diag}\left(\frac{1}{\sigma_i}\right)$$

and, as $\text{diag}(\frac{1}{\sigma_i})^{-1} = \text{diag}(\sigma_i)$, we get

$$\Sigma = \text{diag}(\sigma_i)\Lambda\text{diag}(\sigma_i),$$

which proves the result. $\square$

Corollary 3.2.1 *Σ is invertible if and only if Λ is invertible.*

Proof This results from the fact that in Eq. (3.2.1) $\text{diag}(\sigma_i)$ is invertible, as the risky assets satisfy $\sigma_i \neq 0$. $\square$

To conclude, we demonstrate the property, which we will use later, that if Σ is invertible then two distinct risky assets i and j cannot have a correlation $\rho_{i,j}$ of 1 or -1.

Property 3.2.2 Σ invertible $\implies \left[\forall i, j \in [\![1, d]\!], i \neq j \implies (\rho_{i,j} \neq 1 \text{ and } \rho_{i,j} \neq -1)\right]$.

Proof (by Contradiction).

If we assume that $\rho_{i,j} = 1$ then $\mathbf{Var}(R_i - R_j\frac{\sigma_i}{\sigma_j}) = \sigma_i^2 + \sigma_j^2\frac{\sigma_i^2}{\sigma_j^2} - 2\sigma_i\sigma_j\frac{\sigma_i}{\sigma_j} = 0$, which contradicts that Σ invertible, because if Σ was invertible the only combination of the returns which would give a null variance would be the null combination. Now, if we assume that $\rho_{i,j} = -1$ then $\mathbf{Var}(R_i + R_j\frac{\sigma_i}{\sigma_j}) = \sigma_i^2 + \sigma_j^2\frac{\sigma_i^2}{\sigma_j^2} - 2\sigma_i\sigma_j\frac{\sigma_i}{\sigma_j} = 0$, which contradicts that Σ invertible, as before. $\square$

3.3   Multidimensional Estimations

In the (Markowitz) Gaussian model, the mean and standard deviation parameters for the risky assets can be estimated one by one, as explained in Sect. 1.4.6, or can be estimated together in one go, through **multidimensional estimation** techniques. We present here the construction of these multidimensional estimators, and recall some properties of the **Wishart**, **Hotelling's** T^2, and **Fisher–Snedecor** distributions, which are useful to determine their laws. The proofs are a little complex and require a good background in linear algebra (a good presentation on the topic can be found in Axler [11]), but the results obtained are relatively easily interpretable and implementable.

3.3.1   Wishart, Hotelling's T^2 and Fisher–Snedecor Distributions

The distribution laws for the mean vector and variance-covariance matrix estimators are discussed here. We start with the Wishart distribution, which is the distribution law of the empirical variance-covariance matrix estimator, and which can also be used to determine a **confidence domain** for an estimator of the mean vector, when the variance-covariance matrix is known.

Definition 3.3.1 (Wishart Distribution) If $X_i \sim N(0, \Sigma_d)$ i.i.d. then the law of the random symmetric positive matrix $\sum\limits_{i=1}^{i=n} X_i X_i'$ is called a **Wishart distribution** with n degrees of freedom and is denoted $\mathcal{W}(n, \Sigma_d)$:

$$\sum_{i=1}^{i=n} X_i X_i' \sim \mathcal{W}(n, \Sigma_d).$$

Note that, when $d = 1$, the definition of the Wishart distribution corresponds to the definition of the chi-square distribution. So, the Wishart distribution can be seen as a generalisation, to a matrix space, of the chi-square distribution.

Remark 3.3.1 If Σ_d is invertible, the rank of each linearly matrix $X_i X_i'$ is equal to 1 with probability 1, and the matrices $(X_i X_i')_{i \in [\![1,n]\!]}$ are independent with probability 1 as long as $n \leq d$. From this, one can deduce that, if Σ_d is invertible, $\mathcal{W}(n, \Sigma_d)$ is a probability distribution on the space of symmetric is almost surely invertible positive $d \times d$ matrices of rank $\inf(n, d)$. If $n \geq d$, it can be proven that $\mathcal{W}(n, \Sigma_d)$ has a density distribution $d \times d$ on the **manifold** of dimension $\frac{d(d+1)}{2}$ formed by the symmetric positive definite matrices. In this section, we will manage to prove all the results we need without using the complex calculations related to these density functions. We refer the reader interested in the topic to Anderson [3] and Eaton [39] for more details.

The following properties will be useful when demonstrating the convergence theorems and also help to better understand the nature of the Wishart distribution.

Property 3.3.1 (of the Wishart Distribution)

(1) $\mathcal{W}(n, 1)$ is the chi-squared distribution $\chi^2(n)$.
(2) If $W \sim \mathcal{W}(n, I_d)$ then $\mathbf{tr}(W) \sim \chi^2(nd)$, where $\mathbf{tr}$ is the Trace operator.
(3) If $W \sim \mathcal{W}(n, \Sigma_d)$ and C is a $q \times d$ matrix then $CWC' \sim \mathcal{W}(n, C\Sigma_d C')$.
(4) If $W \sim \mathcal{W}(n, \Sigma_d)$ and $u \in \mathbb{R}^d$ then $u'Wu \sim (u'\Sigma_d u)\chi^2(n)$.
(5) If $W \sim \mathcal{W}(n, \Sigma_d)$ with Σ_d invertible and Z is a random vector of $\mathbb{R}^d$ almost surely not null and independent from W, then

$$\frac{Z'WZ}{Z'\Sigma_d Z} \sim \chi^2(n). \tag{3.3.1}$$

Proof

(1) This results from Definition 1.4.3 of the $\chi^2(n)$ distribution since $\mathcal{W}(n, 1)$ can be expressed as $\sum\limits_{i=1}^{n} X_i^2$ with $X_i \sim \mathcal{N}(0, 1)$ i.i.d.

(2) Let $W = \sum\limits_{i=1}^{n} X_i X_i'$ with $X_i \sim N(0, I_d)$ i.i.d. We write $X_i = (X_i^1, X_i^2, \cdots X_i^d)'$.
     Then $\mathbf{tr}(W) = \sum\limits_{i=1}^{n} \sum\limits_{j=1}^{d} (X_i^j)^2$. As the $(X_i^j)_{i,j \in [\![1,n]\!] \times [\![1,d]\!]}$ are independent $\mathcal{N}(0, 1)$, W is the sum of the squares of $n \times p$ variables $\mathcal{N}(0, 1)$ i.i.d., which proves the result.

(3) Let $W = \sum\limits_{i=1}^{i=n} X_i X_i'$ with $X_i \sim \mathcal{N}(0, \Sigma_d)$ i.i.d., then $CWC' = \sum\limits_{i=1}^{i=n} Y_i Y_i'$ with $Y_i = CX_i$ Gaussian i.i.d. As $\mathbf{Var}(Y_i) = \mathbf{Var}(CX_i) = C\mathbf{Var}(X_i)C' = C\Sigma_d C'$, we conclude that $Y_i \sim \mathcal{N}(0, C\Sigma_d C')$ and the result follows.

(4) is a direct consequence of (3) with $C = u$.

(5) is a generalisation of (4) and we will use (4) to demonstrate it. We use the standard technique of identifying a density function by calculating the expectation of a generic bounded measurable function ϕ, called a test function.

Let ϕ be a bounded measurable function on $\mathbb{R}$. By conditioning on Z we get

$$E\left(\phi\left(\frac{Z'WZ}{Z'\Sigma_d Z}\right)\right) = E\left(E\left[\phi\left(\frac{Z'WZ}{Z'\Sigma_d Z}\right)|Z\right]\right) = \int E\left[\phi\left(\frac{z'Wz}{z'\Sigma_d z}\right)|Z = z\right]f_Z(z)\mathrm{d}z$$

as W is independent of Z and $\frac{z'Wz}{z'\Sigma_d z} \sim \chi^2(n)$ according to (4). If we denote by $f_{\chi^2(n)}(\cdot)$ the density function of a $\chi^2(n)$ distribution, we get

$$E\left[\phi\left(\frac{z'Wz}{z'\Sigma_d z}\right)|Z = z\right] = \int \phi(y)f_{\chi^2(n)}(y)\mathrm{d}y,$$

which is a constant (independent of z), therefore

$$E\left(\phi\left(\frac{Z'WZ}{Z'\Sigma_d Z}\right)\right) = \int\left(\int \phi(y)f_{\chi^2(n)}(y)\mathrm{d}y\right)f_Z(z)\mathrm{d}z = \int \phi(y)f_{\chi^2(n)}(y)\mathrm{d}y,$$

which proves that $\frac{Z'WZ}{Z'\Sigma_d Z}$ has the density of a $\chi^2(n)$ distribution.   □

We continue with the Hotelling's T^2 and Fisher–Snedecor distributions, which are useful when determining a domain of confidence for an estimator of the mean vector, when the variance-covariance matrix is unknown. It also turns out that Hotelling's T^2 and the Fisher–Snedecor distribution are almost the same, as explained in Theorem 3.3.1.

Definition 3.3.2 (Hotelling's T^2 Distribution) Let $X \sim \mathcal{N}(0, \Sigma_d)$ and $W \sim \mathcal{W}(n, \Sigma_d)$ with Σ_d invertible and $n > d - 1$ (for W to be invertible). If we assume that X and W are independent then the law of $nX'W^{-1}X$ is called **Hotelling's T^2 distribution** with parameters d and n and we write

$$nX'W^{-1}X \sim T_d^2(n).$$

Remark 3.3.2 It is easy to check that the law of $X'W^{-1}X$ does not depend on Σ_d and this is the reason why there is no reference to Σ_d in the notation $T_d^2(n)$.

Definition 3.3.3 (Fisher–Snedecor Distribution) If $\chi_1 \sim \chi^2(d_1)$ and $\chi_2 \sim \chi^2(d_2)$ are independent, then

$$\frac{\chi_1/d_1}{\chi_2/d_2}$$

is called a **Fisher–Snedecor law** (or **F-law**) with parameters d_1 and d_2 and is denoted $F(d_1, d_2)$.

To demonstrate the link between the Hotelling's T^2 and Fisher–Snedecor distribution we will need the following two lemmas.

Lemma 3.3.1 *Let $W \sim \mathcal{W}(n, \Sigma_d)$ with Σ_d invertible and $n > d - 1$ (for W to be invertible). Let Z be a random vector of $\mathbb{R}^d$ independent from W and almost surely not null, then*

$$\frac{Z'\Sigma_d^{-1}Z}{Z'W^{-1}Z} \sim \chi^2(n - d + 1).$$

Proof The proof is quite long, even if it only combines arguments of linear projections and of decomposition of the inverse of a block matrix. It can be derived from Eaton [39, Chapter 8, pages 309–313]. □

In the particular case where Z is constant equal to $z \in \mathbb{R}^d \setminus \{0\}$, we obtain

$$\frac{z' \Sigma_d^{-1} z}{z' W^{-1} z} \sim \chi^2(n - d + 1) \tag{3.3.2}$$

and observe that the law obtained does not depend on z.

Lemma 3.3.2 *If $X \sim \mathcal{N}(0, \Sigma_d)$ with Σ_d invertible, then $X' \Sigma_d^{-1} X \sim \chi^2(d)$.*

Proof We can use the **Cholesky decomposition** of a positive definite symmetric matrix $\Sigma_d = \Delta_d \Delta_d'$ and from there $\Sigma_d^{-1} = (\Delta_d')^{-1} \Delta_d^{-1}$. As the transpose and inverse operator commute we have $(\Delta_d')^{-1} = (\Delta_d^{-1})'$. Therefore, $X' \Sigma_d^{-1} X = X'(\Delta_d^{-1})' \Delta_d^{-1} X = (\Delta_d^{-1} X)' \Delta_d^{-1} X$. Now it is easy to show that $\Delta_d^{-1} X \sim \mathcal{N}(0, I_d)$ because

$$\mathbf{Var}(\Delta_d^{-1} X) = \Delta_d^{-1} \mathbf{Var}(X)(\Delta_d^{-1})' = \Delta_d^{-1} \Delta_d \Delta_d'(\Delta_d^{-1})' = I_d$$

and this completes the proof by definition of the chi-square distribution. □

Theorem 3.3.1 (Link Between Fisher–Snedecor and Hotelling's T^2 Distributions) *If $n > d - 1$ then,*

$$T_d^2(n) = \frac{nd}{n - d + 1} F(d, n - d + 1).$$

Proof Using the notations from Eq. (3.3.2)

$$T_d^2(n) \sim n X' W^{-1} X = n\left(\frac{X' W^{-1} X}{X' \Sigma_d^{-1} X}\right) X' \Sigma_d^{-1} X = n \frac{Z_1}{Z_2},$$

with $Z_1 = X' \Sigma_d^{-1} X$ and $Z_2 = \frac{X' \Sigma_d^{-1} X}{X' W^{-1} X}$.

Now, according to Lemmas 3.3.2 and 3.3.1

$$Z_1 \sim \chi^2(d) \text{ and } Z_2 \sim \chi^2(n - d + 1),$$

so

$$T_d^2(n) \sim n \frac{Z_1}{Z_2} \sim n \frac{\chi^2(d)}{\chi^2(n - d + 1)} \sim n \frac{d}{n - d + 1} \frac{\chi^2(d)/d}{\chi^2(n - d + 1)/(n - d + 1)}.$$

Showing that Z_1 and Z_2 are independent will finish the proof. For this, let's consider some measurable bounded (test) functions ϕ and ψ and calculate the expression $E(\phi(Z_1)\psi(Z_2))$:

$$E(\phi(Z_1)\psi(Z_2)) = E\left(E\left[\phi(X'\Sigma_d^{-1}X)\psi(\frac{X'\Sigma_d^{-1}X}{X'W^{-1}X})|X\right]\right)$$

$$= E\left(\phi(X'\Sigma_d^{-1}X)E\left[\psi(\frac{X'\Sigma_d^{-1}X}{X'W^{-1}X})|X\right]\right).$$

Now, $E\left[\psi(\frac{X'\Sigma_d^{-1}X}{X'W^{-1}X})|X = x\right]$ as mentioned in result (3.3.2) is independent of x and therefore can be written $E\left(\psi(\chi^2(n - d + 1))\right)$ or simply $E\left(\psi(Z_2)\right)$. So,

$$E(\phi(Z_1)\psi(Z_2)) = E\left(\phi(X'\Sigma_d^{-1}X)\right)E\left(\psi(Z_2)\right)$$

$$= E\left(\phi(Z_1)\right)E\left(\psi(Z_2)\right).$$

So for all test functions ϕ and ψ, $E(\phi(Z_1)\psi(Z_2)) = E\left(\phi(Z_1)\right)E\left(\psi(Z_2)\right)$ and this proves the independence of Z_1 and Z_2 and thus finishes the proof.   □

3.3.2 Mean Vector and Variance-Covariance Matrix Estimates

When a sample $(X_i)_{i\in[\![1,n]\!]}$ is observed for the distribution $\mathcal{N}(M, \Sigma)$, the **plug-in estimators** for the vector of the expected returns M and the variance-covariance matrix Σ are given as explained in Sect. 1.2 by

$$\widehat{M} = \widehat{E}(X) = \frac{1}{n}\sum_{i=1}^{n} X_i$$

and

$$\widehat{\Sigma}_d = \widehat{\text{Var}}(X) = \frac{1}{n}\sum_{i=1}^{n}(X_i - \widehat{M})(X_i - \widehat{M})'.$$

Note that, for the plug-in estimator of Σ_d, to strictly follow the plug-in method we have used the ratio $\frac{1}{n}$ instead of the more commonly used ratio $\frac{1}{n-1}$ (which is more appropriate to get an unbiased estimator). The various methods to estimate M and Σ_d, and much more about multivariate statistics and hypothesis testing, can be found in Andersen [3] or Rencher [72].

Theorem 3.3.2 (Convergence and Ellipsoidal Confidence Domain) *Let* $(X_i)_{i \in [\![1,n]\!]}$ *be a sample of i.i.d variables of law* $\mathcal{N}(M, \Sigma_d)$ *with* Σ_d *invertible. Let* $\widehat{M} = \bar{X} = \frac{1}{n} \sum_{i=1}^{n} X_i$ *and* $\widehat{\Sigma}_d = \frac{1}{n} \sum_{i=1}^{n} (X_i - \widehat{M})(X_i - \widehat{M})'$*. Then,*

(1) the estimators $\widehat{M}$ *of the mean and* $\widehat{\Sigma}_d$ *of the variance-covariance are independent,*

(2) the estimator $\widehat{M}$ *of the mean satisfies*

$$\sqrt{n}(\widehat{M} - M) \sim \mathcal{N}(0, \Sigma_d),$$

(3) the estimator $\widehat{\Sigma}_d$ *of the variance-covariance matrix satisfies*

$$n\widehat{\Sigma}_d \sim \mathcal{W}(n - 1, \Sigma_d),$$

(4) a confidence domain for $\widehat{M}$*, when the variance-covariance matrix* Σ_d *is known, can be derived from*

$$\sum_{i=1}^{i=n} (X_i - \widehat{M})' \Sigma_d^{-1} (X_i - \widehat{M}) \sim \chi^2((n-1)d),$$

(5) a confidence domain for $\widehat{M}$*, when the variance-covariance matrix* Σ_d *is unknown, can be derived, when* $n > d$*, from*

$$(n-1)(\widehat{M} - M)' \widehat{\Sigma}_d^{-1} (\widehat{M} - M) \sim T_d^2(n-1). \tag{3.3.3}$$

Proof (Hint)
Let $h : \mathbb{R}^{nd} \longrightarrow \mathbb{R}^{(n+1)d}$ be defined by

$$h : X = \begin{pmatrix} X_1 \\ \vdots \\ X_n \end{pmatrix} \longrightarrow \begin{pmatrix} \widehat{M} \\ X_1 - \widehat{M} \\ \cdots \\ X_n - \widehat{M} \end{pmatrix}.$$

Then, X Gaussian and h linear $\Rightarrow h(X)$ is Gaussian. Also, $\forall i$, $\mathbf{Cov}(\widehat{M}, X_i - \widehat{M}) = \frac{1}{n}\mathbf{Cov}(X_i, X_i) - \mathbf{Var}(\widehat{M}) = \frac{1}{n}\Sigma_d - \frac{1}{n^2} \sum_{i=1}^{n} \mathbf{Var}(X_i) = \frac{1}{n}\Sigma_d - \frac{1}{n^2}n\Sigma_d = 0$, so $\widehat{M}$ and $(X_1 - \widehat{M}, \cdots, X_n - \widehat{M})$ are independent, because for Gaussian vectors zero covariance implies independence. As a consequence $\widehat{M}$ and $\widehat{\Sigma}_d$ are independent, which proves (1).

$\sqrt{n}(\widehat{M} - M)$ is the linear transformation of a Gaussian vector and therefore is Gaussian. Now, $\mathbf{E}(\sqrt{n}(\widehat{M} - M)) = \sqrt{n}(\frac{1}{n} \sum_{i=1}^{n} (\mathbf{E}(X_i) - M) = 0$ and $\mathbf{Var}(\sqrt{n}(\widehat{M} - M)) = n \times \frac{1}{n^2} \sum_{i=1}^{n} \mathbf{Var}(X_i) = \Sigma_d$, which proves (2).

For (3) we first consider the case $M = 0$ and $\Sigma_d = \text{Id}$. Let $\otimes$ be the outer product between $\mathbb{R}^d$ and $\mathbb{R}^n$

Let $g : \mathbb{R}^{dn} \longrightarrow \mathbb{R}^d \otimes \mathbb{R}^n$ be defined by

$$g : X = \begin{pmatrix} X_1 \\ \vdots \\ X_n \end{pmatrix} \longrightarrow (X_1 - \widehat{M} | \cdots | X_n - \widehat{M}).$$

$g(X)$ is a random matrix, which is Gaussian, because it is a linear transformation of a Gaussian vector.

Let e_i be the vector of $\mathbb{R}^d$ with all components equal to 0 except for the ith component, which is 1, and let f_n be the vector of $\mathbb{R}^n$ with all components equal to $\frac{1}{\sqrt{n}}$.

Then, $e_i \otimes f_n$ is the $d \times n$ matrix with all components equal to zero, except for the ith row for which all components are equal to $\frac{1}{\sqrt{n}}$.

f_n can be completed with $(f_i)_{i \in [\![1,n-1]\!]}$ to make $(f_i)_{i \in [\![1,n]\!]}$ an orthonormal basis of $\mathbb{R}^n$ and $\mathcal{B} = (e_i \otimes f_j)_{(i,j) \in [\![1,d]\!] \times [\![1,n]\!]}$ a basis of $\mathbb{R}^d \otimes \mathbb{R}^n$.

We now define the **Frobenius scalar product** on $\mathbb{R}^d \otimes \mathbb{R}^n$ by

$$\langle M, N \rangle = \text{Trace}(M N^\perp)$$

It is easy to check that $\mathcal{B}$ is orthonormal with respect to this scalar product and that if $(u_i)_{i \in [\![1,n]\!]}$ and $(v_i)_{i \in [\![1,n]\!]}$ are families of vectors of $\mathbb{R}^d$ and form the vector columns of two matrices then

$$\langle (u_1 | \cdots | u_n), (v_1 | \cdots | v_n) \rangle = \sum_{i=1}^{n} u_i' v_i.$$

Now,

$$\langle g(X), e_i \otimes f_n \rangle = \frac{1}{\sqrt{n}} \sum_{j=1}^{n} \langle X_i - \widehat{M}, e_i \rangle = \frac{1}{\sqrt{n}} \langle \sum_{j=1}^{n} X_i - n\widehat{M}, e_i \rangle = \langle 0, e_i \rangle = 0.$$

So, if we denote by H the space generated by the vectors $\{e_i \otimes f_n, i \in [\![1, d]\!]\}$, $g(X)$ is in $H^\perp$, which is of dimension $(n - 1)d$.

To finish, we just need to prove that $g(X)$ follows a law $\mathcal{N}(0, I_{(n-1)d})$ in $H^\perp$.

Let $u = (u_1| \cdots |u_n)$ be in $H^\perp$, then

$$\mathbf{E}(\langle u, g(X)\rangle) = \langle u, \mathbf{E}(g(X))\rangle = \langle u, 0\rangle = 0$$

and

$$\mathbf{Var}(\langle u, g(X)\rangle) = \mathbf{Var}(\langle u, (X_1|\cdots|X_n) - \sqrt{n}\hat{M} \otimes f_n\rangle)$$

$$= \mathbf{Var}(\langle u, (X_1|\cdots|X_n)\rangle) - \sqrt{n}\langle u, \hat{M} \otimes f_n\rangle).$$

Now, as $\hat{M} \otimes f_n \in H$ and $u \in H^\perp$,

$$\langle u, \hat{M} \otimes f_n\rangle = 0$$

and as the X_i are $\mathcal{N}(0, \mathrm{Id})$ independent, we have

$$\mathbf{Var}(\langle u, X\rangle) = \mathbf{Var}(\sum_{i=1}^{n} u_i' X_i) = \sum_{i=1}^{n} \mathbf{Var}(u_i' X_i) = \|u\|^2.$$

So, finally,

$$\mathbf{Var}(\langle u, g(X)\rangle) = \|u\|^2,$$

which proves that $g(X)$ is a random vector of $H^\perp$ of law $\mathcal{N}(0, I_{(n-1)d})$ and as a consequence can be represented as $g(X) = \sum_{i=1}^{d} \sum_{j=1}^{n-1} Z_{i,j} e_i \otimes f_j$ where $Z_{i,j} \sim \mathcal{N}(0, 1)$ i.i.d. or $g(X) = \sum_{i=1}^{n-1} Z_i \otimes f_i$, where $Z_i \sim \mathcal{N}(0, I_d)$ i.i.d., and therefore

$$n\widehat{\Sigma}_d = g(X)g(X)' = \sum_{i=1}^{n-1} Z_i Z_i' \sim \mathcal{W}(n - 1, I_{d\times d}).$$

In the general case, where $X_i \sim \mathcal{N}(M, \Sigma_d)$, we can write $X_i = \Delta_d Y_i + M$, with $Y_i \sim \mathcal{N}(0, I_d)$ and where Δ_d is obtained by **Cholesky decomposition** of Σ_d and satisfies $\Delta_d \Delta_d' = \Sigma_d$, and we get

$$\sum_{i=1}^{n}(X_i - \widehat{M})(X_i - \widehat{M})' = \Delta_d\Big(\sum_{i=1}^{n}(Y_i - \bar{Y})(Y_i - \bar{Y})'\Big)\Delta_d'$$

$$\sim \Delta_d \mathcal{W}(n - 1, I_{d\times d})\Delta_d'$$

and according to Property 3.3.1

$$\sim \mathcal{W}(n - 1, \Delta_d I_{d\times d}\Delta_d')$$

$$\sim \mathcal{W}(n - 1, \Sigma_d),$$

which proves the result.

(4) is demonstrated from (3) as follows:

According to (3)

$$\sum_{i=1}^{n}(X_i - \widehat{M})(X_i - \widehat{M})' \sim \mathcal{W}(n-1, \Sigma_d),$$

so

$$\Delta_d^{-1}\left(\sum_{i=1}^{n}(X_i - \widehat{M})(X_i - \widehat{M})'\right)\Delta_d^{-1'} \sim \Delta_d^{-1}\mathcal{W}(n-1, \Sigma_d)\Delta_d^{-1'},$$

which, according to Property 3.3.1, has the law of $\mathcal{W}(n-1, \Delta_d^{-1}\Sigma_d\Delta_d^{-1'}) \sim \mathcal{W}(n-1, I_d)$. Applying the trace operator **tr** to each side, we get

$$\mathbf{tr}(\sum_{i=1}^{n}\Delta_d^{-1}(X_i - \widehat{M})(X_i - \widehat{M})'\Delta_d^{-1'}) \sim \mathbf{tr}(\mathcal{W}(n-1, I_d)),$$

which, by the property of commutation under the trace and Property 3.3.1 gives

$$\sum_{i=1}^{n}\mathbf{tr}((X_i - \widehat{M})'\Delta_d^{-1'}\Delta_d^{-1}(X_i - \widehat{M})) \sim \chi^2((n-1)d)$$

as $\Delta_d^{-1'}\Delta_d^{-1} = \Sigma_d^{-1}$ and as the expression under the trace is in fact a number we get the desired result: $\sum_{i=1}^{n}(X_i - \widehat{M})'\Sigma_d^{-1}(X_i - \widehat{M}) \sim \chi^2((n-1)d)$. (5) is a direct consequence of (1), (2) and (3) and of the definition of Hotelling's T^2 distribution $\square$

Corollary 3.3.1 $\frac{n}{n-1}\widehat{\Sigma}_d$ *is an unbiased estimator of* Σ_d.

Proof From Theorem 3.3.2 (3) we have

$$\mathbf{E}(n\widehat{\Sigma}_d) = \mathbf{E}(\mathcal{W}(n-1, \Sigma^d)) = (n-1)\Sigma^d,$$

which proves that $\mathbf{E}\left(\frac{n}{n-1}\widehat{\Sigma}_d\right) = \Sigma^d$.

$\square$

Remark 3.3.3 The mean vector is defined by d parameters. In practice, the number of observations n to estimate the mean vector is a multiple of d, but if not some methods of dimensionality reduction, or shrinkage, linked to machine learning techniques or Bayesian statistics can be considered (see Friedman, Hastie and Tibshirani [48]). If $n > d$, according to Property 3.3.1 Hotelling's T^2 distribution is a Fisher–Snedecor distribution, and Eq. (3.3.3) can be rewritten as

$$(\widehat{M} - M)'\widehat{\Sigma}_d^{-1}(\widehat{M} - M) \sim \frac{d}{n-d}F(d, n-d). \tag{3.3.4}$$

Definition 3.3.4 (Inverted Wishart Distribution) If $\Gamma \sim \mathcal{W}(n, \Sigma_d)$ and $n > d-1$ then Γ is almost surely invertible and then the law of Γ^{-1} is called an **inverted Wishart distribution** and is denoted $\mathcal{W}^{-1}(n, \Sigma_d^{-1})$.

Theorem 3.3.3 *If $n > d + 1$ then,*

$$\mathbf{E}\left(\mathcal{W}^{-1}(n, \Sigma_d^{-1})\right) = \frac{\Sigma_d^{-1}}{n - d - 1}.$$

Proof The proof is quite complex. The interested reader can consult von Rosen [93], who proves the result by using the density function of an inverted Wishart distribution. $\square$

Corollary 3.3.2 *If $n > d + 2$ then*

$$\frac{n - d - 2}{n}\widehat{\Sigma}_d^{-1} \text{ is an unbiased estimator of } \Sigma_d^{-1}.$$

Proof From Theorem 3.3.2 (3), $n\widehat{\Sigma}_d \sim \mathcal{W}(n - 1, \Sigma_d)$, so

$$\frac{1}{n}\widehat{\Sigma}_d^{-1} \sim \mathcal{W}^{-1}(n - 1, \Sigma_d^{-1}).$$

Now, using Theorem 3.3.3

$$\mathbf{E}(\frac{1}{n}\widehat{\Sigma}_d^{-1}) = \mathbf{E}(\mathcal{W}^{-1}(n - 1, \Sigma_d^{-1})) = \frac{1}{n - d - 2}\Sigma_d^{-1}$$

and therefore

$$\mathbf{E}(\frac{n - d - 2}{n}\widehat{\Sigma}_d^{-1}) = \Sigma_d^{-1}.$$

                                                                                          $\square$

3.3.3   Confidence Domain and Statistical Tests

A domain of confidence for the estimator $\widehat{M}$ of the vector M, of the asset's expected returns, can be built. Indeed, if we define

$$\mathcal{D}_q(\widehat{M}) = \{x \in \mathbb{R}^d, (\widehat{M} - x)'\widehat{\Sigma}_d^{-1}(\widehat{M} - x) \leq q\}$$

then, from Eq. (3.3.4),

$$P(M \in \mathcal{D}_q(\widehat{M})) = P\left(\frac{d}{n - d}F(d, n - d) \leq q\right).$$

So, to build from the observations a domain which contains with probability α the true parameter M of the model, one can associate to the estimator $\widehat{M}$ the domain $\mathcal{D}_q(\alpha)(\widehat{M})$, where α is defined by

$$P\left(\frac{d}{n-d}F(d, n-d) \leq q(\alpha)\right) = \alpha.$$

The domain $\mathcal{D}_{q(\alpha)}(\widehat{M})$ is called the **confidence domain** at level α for M, and if we denote by $\Phi_{d,n-d}$ the cumulative distribution function of $F(d, n-d)$, then we have

$$q(\alpha) = \frac{d}{n-d}\Phi_{d,n-d}^{-1}(\alpha). \tag{3.3.5}$$

Finally, when **testing the hypothesis** $H_0 : M = M_0$ at confidence level α, (against $H_1 : M \neq M_0$), H_0 will be accepted if M_0 belongs to $\mathcal{D}_{q(\alpha)}(\widehat{M})$ and rejected otherwise.

3.4   Maket Data with Python

An example is worked out here in dimension 2, for two stocks. Here, the domain $\mathcal{D}_{q(\alpha)}(\widehat{M})$ is an ellipse, which can be easily visualised, while in higher dimensions it is an ellipsoid, more difficult to represent.

The two stocks considered are Deutsche Post (DPW.DE) and Allianz (ALV.VE) and the period of study is the year 2017. The returns are calculated from the daily closing prices of the stocks, and from there the estimators for the vector of expected returns and the variance-covariance matrix are calculated. The estimators are also renormalised on an **annualised basis**, to make things more interpretable. The renormalisation is done in the following way:

If n is the number of returns calculated, between the initial observation date and the final observation date, and if N is the number of days between these two dates, the fraction of time δ between two observations is considered to be (constant) equal to $\frac{N/365}{n}$. So, from a mathematical point of view, the time between a Friday and a Monday (one business day) is considered to be the same as the time between a Monday and a Tuesday. This approach, to measure time on a business day basis, rather than on a calendar day basis, is often discussed but will not be debated here. The estimator for the vector of expected returns $\widehat{M}$, derived from the calculated daily returns $X_i = \frac{P_{i+1} - P_i}{P_i}$, is annualised by multiplying it by $\frac{1}{\delta}$ and the same is done for the estimator of the variance-covariance matrix $\widehat{\Sigma}_d$.

Here, for the estimation of the returns, expressed on an annualised basis, we get

$$\widehat{M} \times \frac{1}{\delta} = (24.63\%, 20.36\%)'$$

and for the estimation of the variance-covariance matrix, expressed on an annualised basis

$$\widehat{\Sigma_d} \times \frac{1}{\delta} = \begin{pmatrix} 0.0242 & 0.0111 \\ 0.0111 & 0.0164 \end{pmatrix},$$

which leads to an annualised standard deviation of the returns of $\sqrt{0.0242} = 15.56\%$ for DPW.DE and $\sqrt{0.0164} = 12.80\%$ for ALV.DE with a correlation between the returns of the two stocks equal to $\rho = \frac{0.0111}{0.1556 \times 0.1280} = 55.73\%$.

We also build a confidence domain, at level 95%, for the vector of the expected daily returns. Here $n = 253$ and $d = 2$, so using the inverse of the Fisher cumulative distribution function $F(2, 251)$ given by Python, we get using Eq. (3.3.5)

$$q(0.95\%) = \frac{2}{253 - 2} \Phi_{2,253-2}^{-1}(0.95) = 0.00796813 \times 3.0316 = 0.02416.$$

In Appendix B.1.1 we show that the ellipse $\mathcal{E}_{q(\alpha)}$ limiting the domain $\mathcal{D}_{q(\alpha)}$ can be parametrised as

$$\mathcal{E}_{q(\alpha)} = \{\widehat{M} + \cos(\theta)\sqrt{\lambda_1 q(\alpha)}u_1 + \sin(\theta)\sqrt{\lambda_2 q(\alpha)}u_2, \theta \in [0, 2\pi].\}$$

The Python program is given in Listing 3.1 and Fig. 3.1 represents, in red, the **confidence ellipse** at level 95% for M. This ellipse is small as the accuracy of the estimator is significant for 253 observations. This confidence ellipse for M is not to be mistaken for the large green ellipse which represents a confidence domain at level 95% for a vector R which follows a Gaussian law $\mathcal{N}(M, \Sigma_d)$, with values for M and Σ_d taken equal to the sample $\widehat{M}$ and $\widehat{\Sigma}_d$. The parametric equation for the large green ellipse is calculated in Appendix B.2. In Fig. 3.2 the confidence ellipse for M is shown at a bigger scale.

Listing 3.1 Python. Data extraction for the Dax Index

```
1  # Determination of the confidence domains
2  # Ref : https://docs.scipy.org/doc/scipy-0.15.1/reference/
        generated/scipy.stats.f.html
3
4  # Library importations
5  import pandas as pd
6  import pandas_datareader.data as web
7  import numpy as np
8  from numpy import linalg as LA
9  from scipy import stats
10 from matplotlib import pyplot as plt
11
12 # Data extraction
13 Tickers = ['DPW.DE','ALV.DE']
14 startinput = '2017-01-01'
15 endinput = '2017-12-31'
```

```
16  alpha_conf = 0.95 # level of confidence for the mean vector
        estimator
17  S = pd.DataFrame() # create the data frame that will contain the
        data
18  for t in Tickers :
19      S[t] = web.DataReader(name = t, data_source='yahoo', start=
            startinput, end= endinput)['Close']
20
21  # Mean vector and variance-covariance matrix calculations
22  R = pd.DataFrame()
23  Covar = pd.DataFrame()
24  Mean = pd.DataFrame()
25  R =S/S.shift(1)-1 # calculate the returns
26  R = R[1:] # eliminate the first raw which is undefined
27  Mean = R.mean() # calculate the mean vector
28  Covar = R.cov() # calculate the variance-covariance matrix
29
30  # Calculation of q(alpha_conf)
31  n = len(R) # calculate the number of returns
32  d = len(Tickers)# calculate the number of stocks used, i.e the
        dimension d.
33  dist= stats.f(d,n-d)
34  [a,b]=dist.interval(2*alpha_conf-1)
35  q = float(d)/float(n-d)*float(b) # calculate q(alpha_conf)
36
37  # Calculation of the average fraction of time between two
        observations
38  Ys= int(startinput[0:4])
39  Ms= int(startinput[5:7])
40  Ds= int(startinput[8:10])
41  Ye= int(endinput[0:4])
42  Me= int(endinput[5:7])
43  De= int(endinput[8:10])
44  import datetime
45  startdate = datetime.date(Ys, Ms, Ds)
46  enddate = datetime.date(Ye, Me, De)
47  z=enddate-startdate
48  Duration = float(z.days) # calculate the number of calendar days
        between the two dates
49  m = float(len(R)-1)
50  delta_t=Duration/365*m # calculate the time interval, delta
51
52  # Calculation of the ellipse characteristics for M
53  w, v = LA.eig(Covar)
54  arg = pd.DataFrame()
55  x = pd.DataFrame()
56  y =  pd.DataFrame()
57  fig, ax = plt.subplots()
58  brg = list(range(100))
59  arg = list(range(100))
60  for i in brg: arg[i]=brg[i]*2*np.pi/99 # list of angles between 0
        and 2 pi
```

```
61  x = np.cos(arg)*np.sqrt(q*w[0])*v.T[0][0]+np.sin(arg)*np.sqrt(q*w
        [1])*v.T[1][0]  +Mean[0]
62  y = np.cos(arg)*np.sqrt(q*w[0])*v.T[0][1]+np.sin(arg)*np.sqrt(q*w
        [1])*v.T[1][1]  +Mean[1]
63  z1 = R['DPW.DE']
64  z2 = R['ALV.DE']
65
66  plt.scatter(z1,z2, marker='+',alpha=0.7)
67  plt.plot(x, y, alpha=0.9,color="red")
68  plt.scatter(Mean[0],Mean[1], color="black", marker='+')
69  plt.xlabel(r'$m_1$')
70  plt.ylabel(r'$m_2$')
71  plt.suptitle('Confidence Domains', size = 'medium')
72  plt.grid()
73
74  # Calculation of the confidence ellipse for a vector of return
75  beta=0.95 # level of confidence
76  Ex = pd.DataFrame()
77  Ey =  pd.DataFrame()
78  distchi = stats.chi2(2)
79  Ea,Eb = dist.interval(2*beta-1)
80  p=Eb
81  Ex = np.cos(arg)*np.sqrt(p*w[0])*v.T[0][0]+np.sin(arg)*np.sqrt(p*
        w[1])*v.T[1][0]  +Mean[0]
82  Ey = np.cos(arg)*np.sqrt(p*w[0])*v.T[0][1]+np.sin(arg)*np.sqrt(p*
        w[1])*v.T[1][1]  +Mean[1]
83  plt.plot(Ex, Ey, alpha=0.7, color="green")
84  # fig.savefig('Confidenceellipse.png', format='png', dpi=1200,
        bbox_inches="tight")
```

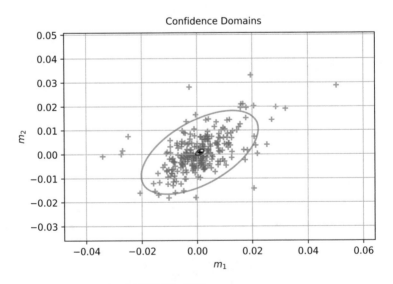

Fig. 3.1 Study for DPW.DE and ALV.DE in 2017

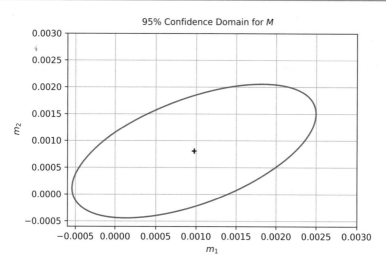

Fig. 3.2 Study for DPW.DE and ALV.DE in 2017

A Few References

1. Amaya, M. P. (2018). *La loi de Wishart*, Université du Québec à Montréal (73 pp.). https://archipel.uqam.ca/12189/1/M15797.pdf.
2. Anderson, T. W. (2003). *An introduction to multivariate statistical analysis* (3rd ed.). New York: Wiley.
3. An Nguyen, T. H, Ruiz-Gazen, A., Thomas-Agnan, C., & Laurent T. (2019). Multivariate student versus multivariate Gaussian regression models with application to finance. *Journal of Risk and Financial Management, 12*(1), 28.
4. Bodnar, T., Mazur, S., & Podgorski, K. (2015). Singular Inverse Wishart Distribution with Application to Portfolio Theory. Department of Statistics, Lund university (Working Papers in Statistics; Nr. 2), 1–17.
5. Eaton, M. L. (2007). *Multivariate statistics: A vector space approach*. IMS Lecture Notes Monograph Series (Vol. 53). https://projecteuclid.org/download/pdf_1/euclid.lnms/1196285114.
6. Korkie, B., & Turtle, H. J. (2002). A mean-variance analysis of self-financing portfolios. *Management Science, 48*(3), 427–444.
7. Markowitz, H. M. (1952, March). Portfolio selection. *Journal of Finance, 7*, 77–91.
8. Markowitz, H. M. (1959). *Portfolio selection: Efficient diversification of investments*. New York: Wiley.
9. Rencher, A. C. (2003). *Methods of multivariate analysis* (2nd ed.). New York: Wiley.

10. Von Rosend, D. (1988). Moments for the inverted Wishart Distribution. *Scandinavian Journal of Statistics, 15*(2), 97–109.
11. Friedman, J., Hastie T., & Tibshirani, R. (2016). *The elements of statistical learning: Data mining, inference, and prediction* (2nd ed.). Springer Series in Statistics. New York: Springer.

Markowitz Without a Risk-Free Asset

4

In this chapter we solve the **Markowitz problem** of finding the investment portfolios which, for a given level of expected return, present the minimum risk. The assumption is made that the returns of the assets (and consequently of the portfolios) follow a Gaussian distribution, and the risk is defined as the standard deviation of the returns. Except in the case where all the risky assets have the same returns, the solution portfolios $\mathcal{F}$ of this mean-variance optimisation problem define a hyperbola when representing in a plane the set $\mathcal{F}(\sigma, m)$ of their standard deviations and expected returns. This hyperbola also determines the limit of all the investment portfolios that can be built. Its upper side $\mathcal{F}^+(\sigma, m)$ corresponds to the efficient portfolios and is called the **efficient frontier**, while its lower side $\mathcal{F}^-(\sigma, m)$ is called the **inefficient frontier**. The **two fund theorem** demonstrated here proves that, when taking any pair of distinct portfolios from $\mathcal{F}$, any other portfolio from $\mathcal{F}$ can be constructed through an allocation between these two portfolios. As a consequence, when two optimal portfolios are found, the subsequent problem of finding other optimal portfolios is just a problem of allocation between these two funds.

4.1 The Optimisation Problem

The d risky assets are assumed to have a vector of returns R which follows a Gaussian law $\mathcal{N}(M, \Sigma)$. We define a new scalar product in $\mathbb{R}^d$ by,

$$\langle x, y \rangle_{\Sigma^{-1}} = x' \Sigma^{-1} y.$$

We define $a = \langle 1_d, 1_d \rangle_{\Sigma^{-1}}$ and $b = \langle M, 1_d \rangle_{\Sigma^{-1}}$ where 1_d is the vector of $\mathbb{R}^d$ with all components equal to 1, and we assume for the time being that $M - \frac{b}{a} 1_d \neq 0$, as the contrary would imply that all the risky assets have the same expected returns and as a consequence that all investment portfolios have the same return as well.

© Springer Nature Switzerland AG 2020
P. Brugière, *Quantitative Portfolio Management*, Springer Texts in Business and Economics, https://doi.org/10.1007/978-3-030-37740-3_4

For an investment portfolio, represented by its allocation π between the d risky assets, the variance of its returns $R(\pi)$ equals

$$\pi'\Sigma\pi,$$

and the expectation of its returns is

$$\pi'M.$$

The fact that the portfolio is an investment portfolio translates into the condition

$$\pi'1_d = 1.$$

Therefore, the optimisation problem to solve here is

$$(P) \begin{cases} \min_{\pi \in \mathbb{R}^d} \pi'\Sigma\pi \\ \pi'M = m \\ \pi'1_d = 1 \end{cases} \tag{4.1.1}$$

which we can write as

$$(P) \begin{cases} \min_{\pi \in \mathbb{R}^d} \langle \Sigma\pi, \Sigma\pi \rangle_{\Sigma^{-1}} \\ \langle \Sigma\pi, M \rangle_{\Sigma^{-1}} = m \\ \langle \Sigma\pi, 1_d \rangle_{\Sigma^{-1}} = 1 \end{cases} \tag{4.1.2}$$

Definition 4.1.1 We denote by $\mathcal{F}$ the set of all the investment portfolio solutions of (P) for any possible value of m and $\mathcal{F}(\sigma, m) = \{(\sigma_\pi, m_\pi), \pi \in \mathcal{F}\}$.

Then we have the following result:

Theorem 4.1.1

$$\mathcal{F} = \{\pi_a + \lambda\omega_{a,b}, \lambda \in \mathbb{R}\} \text{ with } \pi_a = \frac{1}{a}\Sigma^{-1}1_d \text{ and } \omega_{a,b} = \frac{\Sigma^{-1}(M - \frac{b}{a}1_d)}{\|M - \frac{b}{a}1_d\|_{\Sigma^{-1}}}.$$

Proof (P) can be solved either by writing its Lagrangian (see Appendix A.1) or geometrically by noticing that $\Sigma\pi$ must be in the vector space generated by the vectors M and 1_d, denoted $\text{Vect}(M, 1_d)$, in order to mimimise the $\|\cdot\|_{\Sigma^{-1}}$ norm while satisfying the constraints.

As, $\langle M - \frac{b}{a}1_d, 1_d \rangle_{\Sigma^{-1}} = 0$ it appears that $M - \frac{b}{a}1_d$ and 1_d form an orthogonal basis of $\text{Vect}(M, 1_d)$. So, we can search for the solutions π_e of (P) in this basis as

$$\Sigma\pi_e = \lambda 1_d + v(M - \frac{b}{a}1_d),$$

or equivalently

$$\pi_e = \lambda \Sigma^{-1} 1_d + \nu \Sigma^{-1} (M - \frac{b}{a} 1_d).$$

To complete the solution of (P) we observe that the condition $\langle \Sigma \pi_e, 1_d \rangle_{\Sigma^{-1}} = 1$ implies that $\lambda = \frac{1}{a}$ while the condition $\langle \Sigma \pi_e, M \rangle_{\Sigma^{-1}} = m$ determines ν in a unique way as an affine function $\nu(m)$ of m. Therefore the solutions of (P) are the portfolios

$$\pi_e = \frac{1}{a} \Sigma^{-1} 1_d + \nu(m) \Sigma^{-1} (M - \frac{b}{a} 1_d).$$

We can also write these in the form $\pi_a + \lambda(m) \omega_{a,b}$, which finishes the proof. $\square$

In the proposition below we analyse the properties of the portfolios π_a and $\omega_{a,b}$. These properties form the building blocks of the analysis of the properties of any portfolio of $\mathcal{F}$.

Proposition 4.1.1 (Properties of the Portfolios π_a and $\omega_{a,b}$)

(1) $\pi_a' 1_d = 1$ *(so, π_a is an investment portfolio),*
(2) $m_{\pi_a} = \frac{b}{a}$,
(3) $\sigma_{\pi_a} = \frac{1}{\sqrt{a}}$,
(4) $\omega_{a,b}' 1_d = 0$ *(so, $\omega_{a,b}$ is a self-financing portfolio),*
(5) $m_{\omega_{a,b}} = \|M - \frac{b}{a} 1_d\|_{\Sigma^{-1}}$,
(6) $\sigma_{\omega_{a,b}} = 1$,
(7) $\mathbf{Cov}\big(R(\pi_a), R(\omega_{a,b})\big) = 0$.

Proof The proofs are straightforward

(1) $\pi_a' 1_d = 1_d' \pi_a = \frac{1}{a} 1_d' \Sigma^{-1} 1_d = 1$,
(2) $m_{\pi_a} = M' \pi_a = \frac{1}{a} M' \Sigma^{-1} 1_d = \frac{b}{a}$,
(3) $\sigma_{\pi_a} = (\pi_a' \Sigma \pi_a)^{\frac{1}{2}} = (\frac{1}{a^2} 1_d' \Sigma^{-1} \Sigma \Sigma^{-1} 1_d)^{\frac{1}{2}} = (\frac{a}{a^2})^{\frac{1}{2}} = \frac{1}{\sqrt{a}}$,
(4) $\omega_{a,b}' 1_d = \frac{1}{\|M - \frac{b}{a} 1_d\|_{\Sigma^{-1}}} 1_d' \Sigma^{-1} (M - \frac{b}{a} 1_d) = 0$,
(5) $m_{\omega_{a,b}} = M' \omega_{a,b} = M' \frac{\Sigma^{-1} (M - \frac{b}{a} 1_d)}{\|M - \frac{b}{a} 1_d\|_{\Sigma^{-1}}}$, using $1_d' \Sigma^{-1} (M - \frac{b}{a} 1_d) = 0$ we get

$M' \Sigma^{-1} (M - \frac{b}{a} 1_d) = (M - \frac{b}{a} 1_d)' \Sigma^{-1} (M - \frac{b}{a} 1_d)$ and so,

$m_{\omega_{a,b}} = \frac{\|M - \frac{b}{a} 1_d\|_{\Sigma^{-1}}^2}{\|M - \frac{b}{a} 1_d\|_{\Sigma^{-1}}} = \|M - \frac{b}{a} 1_d\|_{\Sigma^{-1}}$,
(6) $(\sigma_{\omega_{a,b}})^2 = \frac{1}{\|M - \frac{b}{a} 1_d\|_{\Sigma^{-1}}^2} (M - \frac{b}{a} 1_d)' \Sigma^{-1} \Sigma \Sigma^{-1} (M - \frac{b}{a} 1_d) = 1$,
(7) $\mathbf{Cov}(R(\pi_a), R(\omega_{a,b})) = \frac{1}{a} 1_d' \Sigma^{-1} \Sigma \Sigma^{-1} (M - \frac{b}{a} 1_d) \frac{1}{\|M - \frac{b}{a} 1_d\|_{\Sigma^{-1}}^2} = 0$. $\square$

From the properties satisfied by the portfolios π_a and $\omega_{a,b}$ we can deduce properties satisfied by the portfolio solutions of (P) which define the frontier $\mathcal{F}$.

Property 4.1.1

(1) $R(\pi_a + \lambda \omega_{a,b}) = R(\pi_a) + \lambda R(\omega_{a,b})$,
(2) $m_{\pi_a + \lambda \omega_{a,b}} = m_{\pi_a} + \lambda m_{\omega_{a,b}}$,
(3) $(\sigma_{\pi_a + \lambda \omega_{a,b}})^2 = (\sigma_{\pi_a})^2 + \lambda^2$,
(4) $\min_{\pi \in \mathcal{F}} \sigma_\pi = \sigma_{\pi_a}$.

Proof

(1) $R(\pi_a + \lambda \omega_{a,b}) = (\pi_a + \lambda \omega_{a,b})' R = \pi_a' R + \lambda \omega_{a,b}' R = R(\pi_a) + \lambda R(\omega_{a,b})$.
(2) follows from (1) by taking expectations on both sides of the equality.
(3) $(\sigma_{\pi_a + \lambda \omega_{a,b}})^2 = \mathbf{Var}\big(R(\pi_a + \lambda \omega_{a,b})\big) = \mathbf{Var}\big(R(\pi_a) + \lambda R(\omega_{a,b})\big)$. As the covariance between these two variables is zero, the expression is $\mathbf{Var}\big(R(\pi_a)\big) + \lambda^2 \mathbf{Var}\big(R(\omega_{a,b})\big)$ and as seen previously $\mathbf{Var}\big(R(\omega_{a,b})\big) = 1$.
(4) This results from (3).   □

4.2   The Geometric Nature of the Set $\mathcal{F}(\sigma, m)$

From the properties satisfied by a portfolio $\pi_e = \pi_a + \lambda \omega_{a,b}$ mentioned above we can define a relationship between m_{π_e} and σ_{π_e} through the parameter λ and therefore describe the geometric nature of $\mathcal{F}(\sigma, m)$. In the first instance we get a parametric representation (with parameter λ) of $\mathcal{F}(\sigma, m)$ and we can also get a cartesian relation by eliminating λ. The results are as follows:

Corollary 4.2.1 *If $\pi_e \in \mathcal{F}$:*

(1) $\pi_e = \pi_a + \frac{m_{\pi_e} - m_{\pi_a}}{m_{\omega_{a,b}}} \omega_{a,b}$,

(2) $(\sigma_{\pi_e})^2 = (\sigma_{\pi_a})^2 + (\frac{m_{\pi_e} - m_{\pi_a}}{m_{\omega_{a,b}}})^2$,

(3) $m_{\pi_e} = m_{\pi_a} + m_{\omega_{a,b}} \sqrt{(\sigma_{\pi_e})^2 - (\sigma_{\pi_a})^2}$ *if $m_{\pi_e} > m_{\pi_a}$,*

(4) $m_{\pi_e} = m_{\pi_a} - m_{\omega_{a,b}} \sqrt{(\sigma_{\pi_e})^2 - (\sigma_{\pi_a})^2}$ *if $m_{\pi_e} < m_{\pi_a}$.*

Proof Trivial when writing π_e as $\pi_a + \lambda \omega_{a,b}$ and eliminating λ between the equations.   □

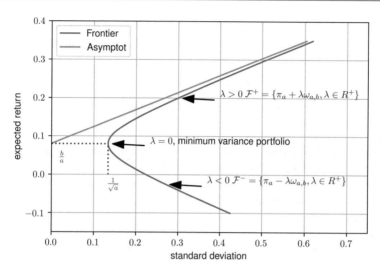

Fig. 4.1 Graph for $\mathcal{F}(\sigma, m)$

Figure 4.1 represents $\mathcal{F}(\sigma, m)$.

Definition 4.2.1 (Efficient and Inefficient Frontier) We define:

- $\mathcal{F}^+ = \left\{ \pi_a + \lambda \omega_{a,b}, \lambda \geq 0 \right\}$,
- $\mathcal{F}^- = \left\{ \pi_a + \lambda \omega_{a,b}, \lambda \leq 0 \right\}$,
- $\mathcal{F}^+(\sigma, m) = \left\{ (\sigma_\pi, m_\pi), \pi \in \mathcal{F}^+ \right\}$ and call it the **Efficient Frontier**,
- $\mathcal{F}^-(\sigma, m) = \left\{ (\sigma_\pi, m_\pi), \pi \in \mathcal{F}^- \right\}$ and call it the **Inefficient Frontier**.

Remark 4.2.1 (Geometric Properties of $\mathcal{F}(\sigma, m)$)

- $\mathcal{F}(\sigma, m)$ is a hyperbola.
- $m = \frac{b}{a}$ is an axis of symmetry for $\mathcal{F}(\sigma, m)$.
- For each $\sigma > \sigma_{\pi_a}$ there are two corresponding points (σ, m_1) and (σ, m_2) on $\mathcal{F}(\sigma, m)$. The point with the higher expected return corresponds to an efficient portfolio and therefore belongs to $\mathcal{F}^+(\sigma, m)$, whereas the other point corresponds to the investment portfolio with the worst possible expected return for this risk and therefore belongs to $\mathcal{F}^-(\sigma, m)$. If (σ, m_1) corresponds to the portfolio $\pi_a + \lambda \omega_{a,b}$ then $\lambda > 0$ and (σ, m_2) corresponds to the portfolio $\pi_a - \lambda \omega_{a,b}$ because $m_{\omega_{a,b}} > 0$.
- For all assets and investment portfolios (σ, m) is on or inside the hyperbola $\mathcal{F}(\sigma, m)$.

Remark 4.2.2 When considering the hyperbola $y = y_0 + \beta\sqrt{x^2 - x_0^2}$ with $\beta > 0$ the line $y = y_0 + \beta x$ is above the hyperbola and is "asymptotically tangent" as:

(1) $y_0 + \beta x - (y_0 + \beta\sqrt{x^2 - x_0^2}) \xrightarrow{x \to \infty} 0$ (points of the curves converging),

(2) $\dfrac{\beta}{\frac{\beta x}{\sqrt{x^2 - x_0^2}}} \xrightarrow{x \to \infty} 1$ (slopes of the curves converging).

Property 4.2.1 (Asymptotic Tangent to $\mathcal{F}(\sigma, m)$) The line intersecting the m-axis at the point $(0, \frac{b}{a})'$ and with slope $m_{\omega_{a,b}} = \|M - \frac{b}{a}1_d\|_{\Sigma^{-1}}$ is asymptotically tangent to $\mathcal{F}(\sigma, m)$. We can also define a cone tangent to $\mathcal{F}(\sigma, m)$ by adding to the first line an additional line intersecting the m-axis at the point $(0, \frac{b}{a})'$ but this time with negative slope $-m_{\omega_{a,b}}$.

Proof This follows directly from the previous remark and corollary.   □

4.3   The Two Fund Theorem

According to Theorem 4.1.1, the portfolio solutions of the optimisation problem (P) consist of an investment in the portfolio π_a combined with an exposure to the self-financing portfolio $\omega_{a,b}$. We can show purely algebraically that such portfolios can also be obtained by an asset allocation between any two distinct portfolios of $\mathcal{F}$.

Proposition 4.3.1 (Two Fund Theorem) *All the portfolios of $\mathcal{F}$ can be built through an asset allocation between any two fixed distinct investment portfolios π_1 and π_2 of $\mathcal{F}$.*

Proof Let $\pi_1 = \pi_a + \lambda_1\omega_{a,b}$ and $\pi_2 = \pi_a + \lambda_2\omega_{a,b}$ be two distinct portfolios of $\mathcal{F}$. Let $\pi = \pi_a + \lambda\omega_{a,b}$ be a portfolio of $\mathcal{F}$. For any $\alpha \in \mathbb{R}$, $\alpha\pi_1 + (1-\alpha)\pi_2$ is an investment portfolio (as $\alpha\pi_1'1_d + (1 - \alpha)\pi_2'1_d = 1$) and if we choose $\alpha = \frac{\lambda - \lambda_2}{\lambda_1 - \lambda_2}$ we get $\alpha\pi_1 + (1 - \alpha)\pi_2 = \pi_a + \lambda\omega_{a,b} = \pi$.   □

Remark 4.3.1 As a consequence, to analyse the portfolios of $\mathcal{F}$ we just need to analyse the combinations of two distinct portfolios of $\mathcal{F}$.

4.3.1   Example with Two Assets: Importance of the Correlation

Let S_1, S_2 be two risky assets with the following characteristics:

$$m_1 = 5\%, \quad \sigma_1 = 15\%,$$
$$m_2 = 20\%, \quad \sigma_2 = 30\%.$$

Let ρ be the correlation between R_1 and R_2.

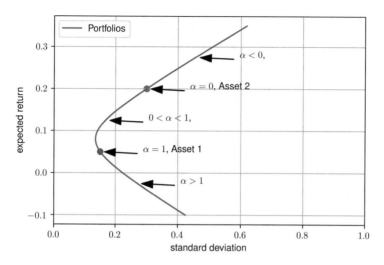

Fig. 4.2 For $\rho = 0$, Σ is invertible and the "usual" hyperbola is obtained

We plot $\mathcal{F}(\sigma, m)$ for $\rho = -1$, $\rho = 0$ and $\rho = 1$ (Figs. 4.2, 4.3 and 4.4).

- The cases $\rho = -1$ and $\rho = 1$ in fact do not correspond to the assumptions made on the risky assets in this section, as in these cases the matrix Σ is not invertible. As a result, a risk-free asset can be built with the two risky assets and the problem of combining two risky assets here becomes equivalent to the problem of combining one risk-free asset and a risk-free asset. As a consequence, in this case the representation of the constructible investment portfolios forms a cone.
- The case $\rho = 0.5$ leads to a matrix Σ which is invertible and in this case we obtain the expected shape for $\mathcal{F}(\sigma, m)$, which is a hyperbola. We call π_α the portfolio for which αx_0 is invested in asset 1 and $(1 - \alpha)x_0$ is invested in asset 2. We represent in the figures in which regions of $\mathcal{F}(\sigma, m)$ the $(\sigma_{\pi_\alpha}, m_{\pi_\alpha})$ of the portfolios π_α stands in the following cases: $\alpha < 0$, $\alpha \in [0, 1]$ and $\alpha > 1$.

4.4   Alternative Parametrisation of $\mathcal{F}(\sigma, m)$ and Conclusion

We derive here another possible parametrisation for the investment portfolios of $\mathcal{F}$. This parametrisation will be useful when determining the "tangent portfolio" in the next section.

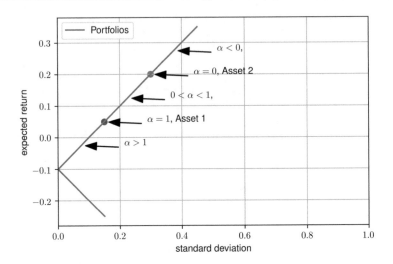

Fig. 4.3 For $\rho = 100\%$, Σ is not invertible and a risk-free asset can be constructed

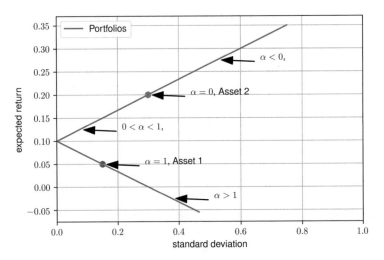

Fig. 4.4 For $\rho = -100\%$, Σ is not invertible and a risk-free asset can be constructed

Proposition 4.4.1 (Alternative Parametrisation of $\mathcal{F}$)

$$\mathcal{F} = \left\{ \frac{1}{b - ma} \Sigma^{-1}(M - m 1_d), m \neq \frac{b}{a} \right\} \bigcup \left\{ \frac{1}{a} \Sigma^{-1} 1_d \right\}.$$

Proof We know that $\mathcal{F} = \left\{ \frac{1}{a} \Sigma^{-1} 1_d + \lambda \Sigma^{-1}(M - \frac{b}{a} 1_d), \lambda \in R \right\}$. For $\lambda = 0$ we obtain the portfolio $\frac{1}{a} \Sigma^{-1} 1_d$. For $\lambda \neq 0$ we can write λ in the form $\lambda = \frac{1}{b - ma}$ with

$m \in \mathbb{R}$ and by doing so we obtain

$$\frac{1}{a}\Sigma^{-1}1_d + \frac{1}{b - ma}\Sigma^{-1}(M - \frac{b}{a}1_d)$$

$$= \frac{1}{a}\Sigma^{-1}1_d + \frac{1}{b - ma}\Sigma^{-1}(M - m1_d) + \frac{1}{b - ma}\Sigma^{-1}(m - \frac{b}{a})1_d$$

$$= \frac{1}{a}\Sigma^{-1}1_d + \frac{1}{b - ma}\Sigma^{-1}(am - b)\frac{1}{a}1_d + \frac{1}{b - ma}\Sigma^{-1}(M - m1_d)$$

$$= \frac{1}{b - ma}\Sigma^{-1}(M - m1_d).$$

$\square$

Remark 4.4.1 We will demonstrate later that the parameter m can be interpreted geometrically, by showing that the tangent to $\mathcal{F}$ at the point (σ, m) intersects the axis $\{\sigma = 0\}$ at the point $(0, m)$.

Remark 4.4.2 We have assumed so far that $M - \frac{b}{a}1_d \neq 0$. We analyse here what would happen if this was not the case. If $M - \frac{b}{a}1_d = 0$ then all the investment portfolios have the same returns equal to $\frac{b}{a}$. In this case, (P) is the problem of minimising, for an investment portfolio, the standard deviation of its return, i.e. to solve:

$$(P) \begin{cases} \min_{\pi}\langle\Sigma\pi, \Sigma\pi\rangle_{\Sigma^{-1}} \\ \langle\Sigma\pi, 1_d\rangle_{\Sigma^{-1}} = 1. \end{cases}$$

As before, from a geometric argument, the solution must satisfy $\Sigma\pi_e \in \text{Vect}(1_d)$, so $\pi_e = \lambda\Sigma^{-1}1_d$. The only π_e of this form satisfying $(\pi_e)'1_d = 1$ is $\pi_e = \frac{1}{a}\Sigma^{-1}1_d$. So, in this case, all the investment portfolios that can be built are on a horizontal half-line in the (σ, m) representation, and the only efficient portfolio is the one of minimal variance π_a, which is at the extremity of this line.

A Few References

1. Markowitz, H. M., Peter, T. G., & Sharpe, W. F. (2000). *Mean-variance analysis in portfolio choice and capital markets*. New York: Wiley.
2. Sharpe, W. F. (1967, March). A linear programming algorithm for mutual fund portfolio selection. *Management Science, 13*(7) Series A, Sciences, 499–510.
3. Sharpe, W. F. (1970). *Portfolio theory and capital markets*. New York: McGraw-Hill.
4. Sharpe, W. F. (1994). The Sharpe ratio. *Journal of Portfolio Management, 21*(1), 49–58.

Markowitz with a Risk-Free Asset

5

In this chapter, a risk-free asset is added to the set of investable securities and the optimal portfolios are now derived in this augmented economy. One of the results obtained is that, in the risk/return analysis, the parameters (σ, m) of any investment portfolio lay either on or inside a certain cone $\mathcal{C}(\sigma, m)$. The upper side of the cone represents the efficient investment portfolios and is called the **Capital Market Line**. We show that the portfolios on the Capital Market Line can be built as an allocation between the risk-free asset and a particular investment portfolio, made of risky assets only, called the **Tangent Portfolio**. The Tangent Portfolio appears to define the tangent point between the cone $\mathcal{C}(\sigma, m)$ and the hyperbola $\mathcal{F}(\sigma, m)$ delimiting the (σ, m) of all investment portfolios made of risky assets only. Based on some economic reasoning, the Tangent Portfolio is sometimes assimilated to the **Market Portfolio**, for which the investment in each asset is proportional to its relative market capitalisation. We also show in this chapter that the problem of optimal allocation can be segmented into two steps. First, the investor decides on the risk exposure he is ready to take, secondly, he calculates the allocation to the Tangent Portfolio which gives him this level of risk (the rest of the money being invested in the risk-free asset). This paradigm for investing is known as the **Separation Theorem** of James Tobin (Nobel Prize in Economics in 1981, see Tobin [90]) and shows that whatever the risk appetite is, there is only one way to take risk exposure efficiently.

The second fundamental result in this chapter is the **Security Market Line** theorem, which links for any investment portfolio (efficient or not) its expected return to the portion of its risk correlated to the returns of the Tangent Portfolio. The **beta** of a portfolio measures the part of its risk correlated to the Tangent Portfolio, and the Security Market Line theorem establishes the affine relationship between the expected return of an asset and its beta.

These results on the Capital Market Line and the Security Market Line are usually known as the **Capital Asset Pricing Model** (CAPM) and were developed initially by William Sharpe (Nobel Prize in Economics 1990), John Lintner and Jan Mossin. The last section of the chapter addresses, for the more advanced readers,

© Springer Nature Switzerland AG 2020
P. Brugière, *Quantitative Portfolio Management*, Springer Texts in Business and Economics, https://doi.org/10.1007/978-3-030-37740-3_5

the problems of stability and of risk on the parameter estimates. **Bayesian methods** are presented to solve these issues, and the **Black–Litterman** model is presented in this context.

5.1   The Optimisation Problem

As there is now an additional risk-free asset, a portfolio Π is now represented by a vector of $\mathbb{R}^{d+1}$. The first component π^0 represents the portion of the wealth invested in the risk-free asset and the additional n-components, represented by a single vector π of $\mathbb{R}^d$, represent the allocation of the wealth in each of the d-risky assets. We denote a portfolio by

$$\Pi = \begin{pmatrix} \pi^0 \\ \pi \end{pmatrix}$$

and the risk-free asset of return r_0 therefore corresponds to

$$\Pi_0 = \begin{pmatrix} 1 \\ 0 \end{pmatrix}.$$

For the time being we assume that $r_0 \neq \frac{b}{a}$, where a and b are defined as before by $a = 1'_d \Sigma^{-1} 1_d$ and $b = 1'_d \Sigma^{-1} M$, and we will study this case later on.

We denote by $R(\Pi)$ the return of the portfolio of allocation Π, $m(\Pi)$ its expected return and $\sigma(\Pi)$ its standard deviation. For a portfolio Π_P we also denote by R_P its return, m_P its expected return and σ_P its standard deviation. Note that, to the risky asset allocation π corresponds only two portfolios: the investment portfolio of risk-free allocation $1 - \pi' 1_d$ and the self-financing portfolio of risk-free allocation $-\pi' 1_d$. Therefore, when we know the nature of a portfolio (investment or self-financing) we can represent it simply by its risky allocation π.

Remark 5.1.1 For an investment portfolio of risky allocation π

- the return is $r_0 + \pi'(R - r_0 1_d)$,
- the expected return is $r_0 + \pi'(M - r_0 1_d)$,
- the variance of the return is $\pi' \Sigma \pi$.

For a self-financing portfolio of risky allocation π

- the return is $\pi'(R - r_0 1_d)$,
- the expected return is $\pi'(M - r_0 1_d)$,
- the variance of the return is $\pi' \Sigma \pi$.

The efficient investment portfolios, which are indeed characterised by their risky allocation π (with $\pi^0 = 1 - \pi'1_d$), are now the solutions of

$$(Q) \begin{cases} \min_{\pi \in \mathbb{R}^d} \pi' \Sigma \pi \\ r_0 + \pi'(M - r_0 1_d) = m \end{cases}$$

or equivalently

$$(Q) \begin{cases} \min_{\pi \in \mathbb{R}^d} \langle \Sigma \pi, \Sigma \pi \rangle_{\Sigma^{-1}} \\ \langle \Sigma \pi, M - r_0 1_d \rangle_{\Sigma^{-1}} = m - r_0. \end{cases}$$

5.2 Capital Market Line and Limit Cone $\mathcal{C}(\sigma, m)$

Notations Let

$$\Pi_T = \begin{pmatrix} 0 \\ \pi_T \end{pmatrix} \text{ with } \pi_T = \frac{1}{b - r_0 a} \Sigma^{-1}(M - r_0 1_d).$$

Property 5.2.1 (Analysis of Π_T)

(1) Π_T represents an investment portfolio with no allocation to the risk-free asset.
(2) $\mathbf{E}(R(\Pi_T)) = r_0 + \frac{1}{b - r_0 a} \|M - r_0 1_d\|_{\Sigma^{-1}}^2$.
(3) $\mathbf{Var}(R(\Pi_T)) = \frac{1}{(b - r_0 a)^2} \|M - r_0 1_d\|_{\Sigma^{-1}}^2$.

Proof

For (1)

$$1_d' \frac{1}{b - r_0 a} \Sigma^{-1}(M - r_0 1_d) = \frac{1}{b - r_0 a}(b - r_0 a) = 1.$$

For (2)

$$\mathbf{E}(R(\Pi_T)) = r_0 + (M - r_0 1_d)' \pi_T$$

$$= r_0 + (M - r_0 1_d)' \frac{1}{b - r_0 a} \Sigma^{-1}(M - r_0 1_d)$$

$$= r_0 + \frac{\|M - r_0 1_d\|_{\Sigma^{-1}}^2}{b - r_0 a}.$$

For (3)

$$\mathbf{Var}(R(\Pi_T)) = \pi_T' \Sigma \pi_T$$

$$= \frac{1}{(b - r_0 a)^2} (M - r_0 1_d)' \Sigma' \Sigma^{-1} \Sigma (M - r_0 1_d)$$

$$= \frac{1}{(b - r_0 a)^2} \|M - r_0 1_d\|_{\Sigma^{-1}}^2 \text{ as } \Sigma' = \Sigma,$$

which proves the result. $\square$

Theorem 5.2.1 (Separation Theorem) *The investment portfolio solutions of (Q) have for allocations*

$$\beta \Pi_T + (1 - \beta) \Pi_0$$

with

$$\beta = \frac{m - r_0}{m_T - r_0}.$$

Proof (Q) can be solved geometrically like (P) in (4.1.2). By geometrical arguments the solutions of (Q) are of the form $\Sigma \pi = \lambda(M - r_0 1_d)$, or equivalently, $\pi = \lambda \Sigma^{-1}(M - r_0 1_d)$. Therefore, an investment portfolio solution Π of (Q) has a risky allocation of the form $\beta \pi_T$ and a risk-free allocation equal to $1 - 1_d'(\beta \pi_T) = 1 - \beta$. As a consequence, such a portfolio is of the form $\beta \Pi_T + (1 - \beta)\Pi_0$, where the parameter β is identified by the condition on the mean

$$r_0 + \pi'(M - r_0 1_d) = m$$

$$\Longleftrightarrow \beta m_T + (1 - \beta)r_0 = m$$

and as $m_T = r_0 + \frac{1}{b - r_0 a} \|M - r_0 1_d\|_{\Sigma^{-1}}^2 > r_0$

$$\Longleftrightarrow \beta = \frac{m - r_0}{m_T - r_0},$$

which proves the result. $\square$

Remark 5.2.1 The investment portfolio solutions of (Q) are built by allocating money between only two portfolios Π_0 and Π_T. Once Π_T is determined, the problem of building an optimal portfolio is then reduced to the problem of choosing the percentage β of the wealth invested in Π_T. The portfolio Π_T represents the most efficient way to take risk.

Notations Let

$$\mathcal{C} = \{\lambda \Pi_T + (1 - \lambda)\Pi_0, \lambda \in \mathbb{R}\}$$

be the set of all investment portfolio solutions of (Q) and

$$\mathcal{C}(\sigma, m) = \{(\sigma(\Pi), m(\Pi)), \Pi \in \mathcal{C}\}$$

be the set they form in the risk/return representation.

Property 5.2.2 (Geometric Nature of $\mathcal{C}(\sigma, m)$) Let $\Pi_\lambda = \lambda \Pi_T + (1 - \lambda)\Pi_0$ be a portfolio of $\mathcal{C}$ and let $m_\lambda = m(\Pi_\lambda)$ and $\sigma_\lambda = \sigma(\Pi_\lambda)$. Then,

(1) $m_\lambda = \lambda m(\Pi_T) + (1 - \lambda)m(\Pi_0)$,
(2) $\sigma_\lambda = |\lambda|\sigma(\Pi_T)$,
(3) $\mathcal{C}(\sigma, m)$ is a cone and
(4) for any other investment portfolio Π, $(\sigma(\Pi), m(\Pi))$ is inside the cone $\mathcal{C}(\sigma, m)$.

Proof

(1) $\mathbf{E}(R_\lambda) = \lambda \mathbf{E}(R(\Pi_T)) + (1 - \lambda)\mathbf{E}(R(\Pi_0)) = \lambda m(\Pi_T) + (1 - \lambda)m(\Pi_0)$.
(2) $\mathbf{Var}(R_\lambda) = \mathbf{Var}(\lambda R(\Pi_T) + (1 - \lambda)R(\Pi_0)) = \mathbf{Var}(\lambda R(\Pi_T)) = \lambda^2 \mathbf{Var}(R(\Pi_T))$, which implies that $\sigma_\lambda = |\lambda|\sigma(\Pi_T)$.
(3) Properties (1) and (2) give a parametric representation of a cone with parameter λ.
(4) For any investment portfolio Π, the portfolio solution Π_e of (Q) for $m = m(\Pi)$ satisfies $\sigma(\Pi_e) \leq \sigma(\Pi)$, so the point $(\sigma(\Pi), m(\Pi))$ is to the right of $(\sigma(\Pi_e), m(\Pi_e))$ and as a consequence inside the cone $\mathcal{C}(\sigma, m)$. $\qquad \square$

Definition 5.2.1 (Capital Market Line) The upper side of the cone $\mathcal{C}(\sigma, m)$ is called the **Capital Market Line** or **Efficient Frontier** and is denoted $\mathcal{C}^+(\sigma, m)$. The lower side of the cone is called the **Inefficient Frontier** and is denoted $\mathcal{C}^-(\sigma, m)$ (Fig. 5.1).

5.2.1 The Market Portfolio

In this section we define the **Market Portfolio**, for which the risky allocation in asset i corresponds to the weight that company i has in the economy. We show that, under certain economic assumptions, the Market Portfolio and the portfolio Π_T should be the same.

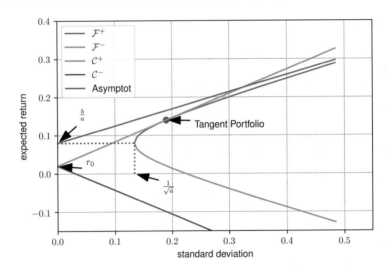

Fig. 5.1 Capital Market Line

Definition 5.2.2 (Market Portfolio) The **Market Portfolio** is the investment portfolio, with no risk-free component, for which the weight of each risky asset i is equal to the market capitalisation of company i divided by the sum of the market capitalisations of all the companies in the economy.

Proposition 5.2.1 *If we assume that:*

(1) all investors have the same model assumptions and allocate efficiently according to the Capital Market Line theorem and
(2) the market value of each company is equal to the total amount invested in this company,

then the portfolio Π_T and the Market Portfolio Π_M are the same.

Proof Let $(x_j)_{j \in [\![1,N]\!]}$ be the initial wealths of the N investors and $(\lambda_j)_{j \in [\![1,N]\!]}$ be the portions of their wealth (depending on their risk appetite) invested in the portfolio Π_T. The total amount invested in the risky asset i is therefore $\sum_{j=1}^{j=N} \lambda_j x_j \pi_T^i$.

According to the assumptions, the market capitalisation MCap_i of company i is equal to the total amount invested in this company and therefore

$$\mathrm{MCap}_i = \sum_{j=1}^{j=N} \lambda_j x_j \pi_T^i.$$

Also, for the total market capitalisation we get

$$\text{TotalMCap} = \sum_{i=1}^{i=n} \text{MCap}_i = \sum_{j=1}^{j=N} \lambda_j x_j.$$

Therefore the weight of company i in the economy (with respect to its market capitalisation), which is its weight in the Market Portfolio, is $\frac{\text{MCap}_i}{\text{TotalMCap}} = \pi_i$, which is the weight in the Portfolio Π_T.   □

In practice, the portfolio Π_T may be different from the Market Portfolio. Proposition 5.2.2 gives a mathematical (and not economical this time) sufficient and necessary condition for the two portfolios to be the same.

Proposition 5.2.2 (Condition of Optimality for the Market Portfolio)   *Let* π_M *be the vector of risky allocations for the Market Portfolio. The Market Portfolio is the portfolio* Π_T *if and only if*

$$M - r_0 1_d \propto \Sigma \pi_M$$

($\propto$ *meaning 'proportional to'*).

Proof   First, we assume that $\pi_T = \pi_M$, then

$$\pi_T = \pi_M \Longrightarrow \frac{1}{b - r_0 a} \Sigma^{-1}(M - r_0 1_d) = \pi_M$$

$$\Longrightarrow M - r_0 1_d = (b - r_0 a) \Sigma \pi_M$$

$$\Longrightarrow M - r_0 1_d \propto \Sigma \pi_M.$$

Second, we assume that $M - r_0 1_d = \alpha \Sigma \pi_M$ with $\alpha \in \mathbb{R}$, then

$$\pi_T = \frac{1}{b - r_0 a} \Sigma^{-1} \Big(\alpha \Sigma \pi_M \Big)$$

$$\Longrightarrow \pi_T = \frac{\alpha}{b - r_0 a} \pi_M$$

with

$$b = 1_d' \Sigma^{-1}(\alpha \Sigma \pi_M + r_0 1_d) = \alpha 1_d' \pi_M + r_0 a = \alpha + r_0 a$$

and consequently,

$$\pi_T = \frac{\alpha}{\alpha + r_0 a - r_0 a} \pi_M = \pi_M,$$

which finishes the proof.   □

Remark 5.2.2 A market-weighted index has by definition the same weights as the Market Portfolio and therefore should be seen, if the economic assumptions of Proposition 5.2.1 are satisfied, as the best instrument to take exposure to risk, in the Markowitz framework.

Remark 5.2.3 This model, with a risk-free asset, leading to the result of the Capital Market Line, is called the **CAPM** (Capital Asset Pricing Model) or sometimes **MEDAF** (*modèle d'évaluation des actifs financiers*).

5.2.2   The Tangent Portfolio

Property 5.2.3 The risky allocation π_T of the portfolio Π_T is a solution of the mean/variance optimisation problem (P) in Eq. (4.1.1) (with no risk-free asset). So, $\pi_T \in \mathcal{F}$.

Proof

$$
\begin{aligned}
\pi_T &= \frac{1}{b - r_0 a} \Sigma^{-1} (M - r_0 1_d) \\
&= \frac{1}{b - r_0 a} \Sigma^{-1} \left((M - \frac{b}{a} 1_d) + (\frac{b}{a} 1_d - r_0 1_d) \right) \\
&= \frac{1}{b - r_0 a} (\frac{b}{a} - r_0) \Sigma^{-1} 1_d + \Sigma^{-1} (M - \frac{b}{a} 1_d) \\
&= \frac{1}{a} \Sigma^{-1} 1_d + \frac{1}{b - r_0 a} \Sigma^{-1} (M - \frac{b}{a} 1_d) \\
&= \pi_a + \lambda \omega_{a,b},
\end{aligned}
$$

which is the expression of the portfolios of $\mathcal{F}$.   □

Remark 5.2.4 Usually in the model $\frac{b}{a} > r_0$, otherwise the expected return of the portfolio Π_T would be less than r_0 and the Capital Market Line would be made of investment portfolios obtained by taking a short position in the portfolio Π_T, which economically does not make sense as it is would mean that there would be only sellers for this position.

For this reason, until the end of this section we assume that $\frac{b}{a} > r_0$, which also means that the axis of symmetry $\{m = \frac{b}{a}\}$ for the hyperbola $\mathcal{F}(\sigma, m)$ is above the line $\{m = r_0\}$.

Property 5.2.4 (Tangent Portfolio) $\mathcal{C}(\sigma, m)$ is tangent to $\mathcal{F}(\sigma, m)$ at the point $(\sigma(\Pi_T), m(\Pi_T))$ and therefore Π_T is called the Tangent Portfolio.

Proof By definition of $\mathcal{C}$, Π_T belongs to $\mathcal{C}$ and from Property 5.2.3 Π_T is in $\mathcal{F}$ as well. Geometrically, if a line and a hyperbola have a contact point then, either they are tangent at this contact point or they cross each other. The situation where they cross each other is not possible here as it would imply that some portfolios of $\mathcal{F}$ are more efficient than any portfolio of $\mathcal{C}$.   □

Remark 5.2.5 Equivalently, we can say that the tangent to the Efficient Frontier $\mathcal{F}(\sigma, m)$ at the point $(\sigma(\Pi_T), m(\Pi_T))$ intersects the $\{\sigma = 0\}$ axis at the point $\begin{pmatrix} 0 \\ r_0 \end{pmatrix}$.

5.2.3   More Geometric Properties

Corollary 5.2.1 (Geometry of the Efficient Frontier)

(1) For any investment portfolio $\pi = \frac{1}{b-ma}\Sigma^{-1}(M - m1_d)$ of $\mathcal{F}$, the tangent to $\mathcal{F}(\sigma, m)$ at $(\sigma(\pi), m(\pi))$ intersects the $\{\sigma = 0\}$ axis at the point $\begin{pmatrix} 0 \\ m \end{pmatrix}$.

(2) For the minimum variance portfolio of $\mathcal{F}$, $\pi_a = \frac{1}{a}\Sigma^{-1}1_d$, the tangent to $\mathcal{F}(\sigma, m)$ at $(\sigma(\pi_a), m(\pi_a))$ is parallel to the $\{\sigma = 0\}$ axis.

Proof

(1) follows from Property 5.2.4 when taking $r_0 = m$.
(2) follows from the properties of the hyperbola $\mathcal{F}(\sigma, m)$.   □

Remark 5.2.6 These results mean that there is a bijection between $\mathcal{F} - \{\frac{1}{a}\Sigma^{-1}1_d\}$ and $\mathbb{R} - \{\frac{b}{a}\}$ and that it is the tangents to the $\mathcal{F}(\sigma, m)$ curve which establish the bijection between the portfolios and their parameters m.

5.3   The Security Market Line

Let $\rho(X, Y)$ be the linear correlation between two random variables X and Y.

Theorem 5.3.1 (Security Market Line) *Let Π_T be the Tangent Portfolio and Π_P be an investment portfolio composed of the risk-free and risky assets, then*

$$m_P - r_0 = (m_T - r_0)\rho(R_P, R_T)\frac{\sigma_P}{\sigma_T} \tag{5.3.1}$$

and

$$R_P - r_0 = (R_T - r_0)\rho(R_P, R_T)\frac{\sigma_P}{\sigma_T} + \epsilon_P, \tag{5.3.2}$$

with ϵ_P normal independent from R_T and centered.

Proof Let Π_P be an investment portfolio then,

$$\mathbf{Cov}(R_T, R_P) = \pi_T' \Sigma \pi_P = \frac{1}{b - r_0 a}(M - r_0 1_d)' \pi_P = \frac{m_P - r_0}{b - r_0 a}.$$

The same calculation applied to Π_T gives $\mathbf{Cov}(R_T, R_T) = \frac{m_T - r_0}{b - r_0 a}$ and from there,

$$\mathbf{Cov}(R_T, R_P) = \frac{m_P - r_0}{m_T - r_0} \mathbf{Cov}(R_T, R_T),$$

which we can write as $m_P - r_0 = (m_T - r_0)\rho(R_T, R_P)\frac{\sigma_P}{\sigma_T}$, proving Eq. (5.3.1).

We now show a relationship for the random variables and not only for their expectations. We observe that

$$\begin{pmatrix} (R_P - r_0) - (R_T - r_0)\rho(R_T, R_P)\frac{\sigma_P}{\sigma_T} \\ R_T \end{pmatrix}$$

is Gaussian because it is an affine transformation of the vector of the returns R of the risky assets, which is assumed to be Gaussian. Therefore, to show that the first component, which we call ϵ_P, is independent of the second one we just need to show that the covariance is zero. Indeed,

$$\mathbf{Cov}(\epsilon_P, R_T) = \mathbf{Cov}(R_P - R_T \rho(R_T, R_P)\frac{\sigma_P}{\sigma_T}, R_T)$$

$$= \mathbf{Cov}(R_P, R_T) - \rho(R_T, R_P)\frac{\sigma_P}{\sigma_T}\mathbf{Cov}(R_T, R_T)$$

$$= 0.$$

The fact that $\mathbf{E}(\epsilon_P) = 0$ results from Eq. (5.3.1) and this finishes the proof.   □

Definition 5.3.1 (Beta of an Asset) Equation 5.3.1 is called the **Security Market Line equation**. The quantity $\rho(R_T, R_P)\frac{\sigma_P}{\sigma_T}$ is denoted $\beta_T(P)$ and is called the **beta** of the portfolio Π_P relative to the portfolio Π_T.

Remark 5.3.1 The equation $m_P - r_0 = (m_T - r_0)\beta_T(P)$

- is valid for all investment portfolios and not only for efficient portfolios,
- shows that only the risk correlated with the Tangent Portfolio is remunerated,
- is used in capital budgeting to determine the price of an asset/a venture based on its expected returns and beta with the sector.

The beta can be estimated by statistical regression of the observed excess returns $R_P - r_0$ over $R_T - r_0$ on several periods.

Exercise 5.3.1 Show that for a self-financing portfolio Q the SML equation (5.3.2) becomes

$$R_Q = (R_T - r_0)\rho(R_Q, R_T)\frac{\sigma_P}{\sigma_T} + \epsilon_Q.$$

Property 5.3.1 From the SML equation $R_P = r_0 + (R_T - r_0)\beta_T(P) + \epsilon_P$ we get:

(1) $\sigma_P^2 = \sigma_T^2 \beta_T(P)^2 + \sigma^2(\epsilon_P)$,
(2) $\sigma_T \beta_T(P) = \rho(R_T, R_P)\sigma_P$,
(3) $\sigma(\epsilon_P) = \sqrt{1 - \rho^2(R_T, R_P)}\sigma_P$.

Proof

(1) results from the independence of R_T and ϵ_P.
(2) $\beta_T(P) = \frac{\mathbf{Cov}(R_T, R_P)}{\sigma_T^2} = \rho(R_T, R_P)\frac{\sigma_T \cdot \sigma_P}{\sigma_T^2} \implies \beta_T(P)\sigma_T = \rho(R_T, R_P)\sigma_P$.
(3) $\sigma^2(\epsilon_P) = \sigma_P^2 - \sigma_T^2 \beta_T(P)^2 = \sigma_P^2 - \rho(R_T, R_P)^2 \sigma_P^2$, which proves the result. $\square$

Definition 5.3.2

- $\sigma_T|\beta_T(P)|$ is called the **systematic risk**. Economically, the fact that it is remunerated is explained by the fact that it cannot be reduced by diversification.
- $\sigma(\epsilon_P)$ is called the **idiosyncratic risk** or **specific risk**. Economically, the fact that it is not remunerated is explained by the fact that a risk that can be reduced by diversification does not need to be remunerated.

Exercise 5.3.2 Let Π_P be an investment portfolio with risk parameters (σ_P, m_P). Let (x, m_P) be the intersection of the cone $\mathcal{C}(\sigma, m)$ and the line $\{m = m_P\}$. Using the SML equation, show that $x = \rho(R_T, R_P)\sigma_P$. Conclude that in a risk/reward representation as in Fig. 5.2 we can read the decomposition between systematic risk and idiosyncratic risk.

5.3.1 The Security Market Line and "Arbitrage" Detections

In theory, the (β, m) of the assets should be aligned according to the SML. In practice, the real parameters of the model are unknown and the fact that we are dealing only with estimates, and that the model is not perfect, produces some points which are not perfectly aligned. A trading idea is to play a reversion of the observed (β, m) points to the estimated Security Market Line. A possible implementation is described below.

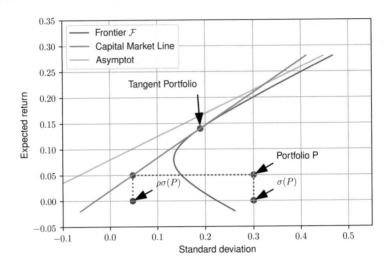

Fig. 5.2 Systematic risk and specific risk

- Take as the Tangent Portfolio the broader index to which the stocks belong.
- Take for the beta of each stock the beta to this broader index, calculated on a historical basis.
- Use for the expected returns of each stock analyst predictions, linked to estimated profits divided by the current share price.

In practice, the points will not be perfectly aligned and a regression line will be calculated.

- The assets over the line will look cheap, as they are expected to revert to the line through a decrease of their expected returns (their beta is assumed to be stable) and as such a decrease is assumed to take place via an increase of the share price, and not a decrease of the expected profits, which are supposed to be relatively stable.
- The assets below the line will look expensive for the opposite reason.

In "pair-trading", strategies will be considered consisting in:

- Selling assets lying below the estimated Security Market Line.
- Buying assets lying above the Security Market Line (usually in the same sector).

5.4   Market Data with Python

5.4.1   The Frontier and Capital Market Line for the DAX 30 Components

First, we limit ourselves to five stocks of the DAX 30, in order to be able to better analyse the results. We calculate for these stocks their average returns and variance-covariance matrix, based on daily observations. Then, we normalise these estimates to express them on an annualised basis. We use zero for the risk-free interest rate. The Python code to produce the Efficient Frontier, the Capital Market Line, the Security Market Line and to calculate the expected returns, standard returns, correlation and variance-covariance matrices is given in Listing 5.1:

Listing 5.1 Python. Risk versus reward for 5 stocks

```
1  # Library importations
2  import pandas as pd
3  import pandas_datareader.data as web
4  import numpy as np
5  import math
6  from numpy import linalg as LA
7  from numpy.linalg import inv
8  from scipy import stats
9  #from matplotlib import pyplot as plt # delete for Python 3
10 import matplotlib.pyplot as plt # add for Python 3
11 fig, ax = plt.subplots()
12
13 # Data extraction
14 data = pd.Series()
15 TickerDax = ['DPW.DE','ALV.DE','BMW.DE','BAS.DE','FME.DE']
16 for x in TickerDax:
17     data[x] = web.DataReader(name = x, data_source='yahoo', start=
       '2017-01-1',end='2017-12-31')
18
19 # Variable creations
20 r=0 # risk-free rate
21 AFactor = len(data['DPW.DE'])-1 # annualisation factor, based on
       number of observations in a year
22 n_TickerDax = len(TickerDax) # nb of stocks considered
23 Vec1 = np.linspace(1, 1, n_TickerDax) # create a vector, with all
       components equal to 1
24 data_R = pd.DataFrame() # daily returns
25 Mdata_R = pd.DataFrame() # means of the daily returns
26 Sigma = pd.DataFrame() # variance-covariance matrix of the
       returns annualised
27 Sigma_diag = pd.DataFrame() # individual variances (derived from
       the variance-covariance matrix)
28
29 for x in TickerDax:
30     data_R[x] = (data[x]['Close']/data[x]['Close'].shift(1)-1)
```

```
31  Sigma = AFactor*data_R.cov() # annualisation, the missing values
        are not taken into account with .Cov()
32  InvSigma = inv(Sigma)
33  for x in TickerDax:
34      Sigma_diag[x] = [Sigma[x][x]]
35      Mdata_R[x]= [float(np.mean(data_R[x])) ]
36
37  # Model parameters, derived from the observations
38  Mean =   AFactor* Mdata_R.iloc[0] # average returns on an
        annualised basis
39  STD = Sigma_diag.iloc[0]**.5 # standard deviations on an
        annualised basis
40
41  # Important variables calculation
42  a = Vec1.T.dot(InvSigma).dot(Vec1)
43  b = Mean.T.dot(InvSigma).dot(Vec1)
44  sd_a = 1 / math.sqrt(a) # standard deviation, minimum variance
        portfolio
45  m_a = b / a # expected return, minimum variance portfolio
46  m_w = math.sqrt((Mean - b/a * Vec1).T.dot(InvSigma).dot(Mean - b/
        a * Vec1)) # expected return, portfolio w
47  m_r = math.sqrt((Mean - r * Vec1).T.dot(InvSigma).dot(Mean - r *
        Vec1)) # expected return, Tangent Portfolio
48
49  # Graph plotting
50  np.random.seed(7777) # Fixing random state for reproducibility
51  colors = np.random.rand(len(STD)) # each point will have its own
        (random) color
52  plt.scatter(STD, Mean, c=colors, alpha=0.7) # plot the assets,
        alpha transparency parameter
53  ax.annotate(TickerDax[0], (STD[0]+.01,Mean[0] ))
54  ax.annotate(TickerDax[1], (STD[1]+.01,Mean[1] ))
55  ax.annotate(TickerDax[2], (STD[2]+.01,Mean[2] ))
56  ax.annotate(TickerDax[3], (STD[3]+.01,Mean[3]-.01 ))
57  ax.annotate(TickerDax[4], (STD[4]+.01,Mean[4] ))
58  range_inf = np.min(Mean) - 0.05
59  range_sup = np.max(Mean) + 0.35
60  z2 = np.linspace(range_inf , range_sup, 50) # range of expected
        returns considered
61  z1 = pd.DataFrame()
62  z= pd.DataFrame()
63  zr= pd.DataFrame()
64  i=0
65  for i in range(len(z2)):
66      z1[i]= [math.sqrt( ((z2[i]- m_a)/m_w)**2 + sd_a**2 )] #
        Frontier
67      z[i]= [(z2[i] - b/a)/m_w] # Asymptot to the Frontier
68      zr[i]= [(z2[i] - r)/m_r] # Capital Marlet Line
69  z1 = z1.iloc[0]
70  z=z.iloc[0]
71  zr=zr.iloc[0]
72  plt.plot(z1, z2, alpha=1) # plot the Frontier
73  plt.plot(z, z2, alpha=1) # plot the asymptot to the Frontier
```

```
74  plt.plot(zr, z2, alpha=1) # plot the Capital Market Line
75  plt.legend(['Frontier','Asymptot','Capital Market Line'], loc=2)
76  plt.grid(True)
77  plt.xlabel('Annualised standard deviation')
78  plt.ylabel('Annualised return')
79
80  # Useful Ouputs.
81  print ("annualised standard deviations of the returns:")
82  print (STD)
83  print ("annualised means of the returns:")
84  print (Mean)
85  print ("min variance portfolio:")
86  print ("sd_a", sd_a, ",m_a", b/a)
87  print ("matrix of variance-covariance:")
88  print (Sigma)
89  print ("Correl Matrix")
90  print (data_R . corr ())
```

In Fig. 5.3 we can visualise the risk/return analysis for these five stocks and in Table 5.1 their risk/return parameters.

For the minimum variance portfolio the calculations give

- an annualised standard deviation of 10.76% and
- an annualised mean of 7.39%.

The variance-covariance and correlation matrices for the returns are calculated in Tables 5.2 and 5.3. Note that the standard deviations in the program are calculated from the variance-covariance matrix, and not directly from the individual returns, in order to avoid any inconsistency that the statistical functions of Python 3.6 may

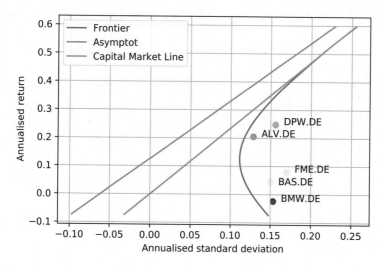

Fig. 5.3 Capital Market Line, 5 stocks

Table 5.1 Mean and
standard deviations of the five
stocks

Stock	Mean	Standard deviation
DPW.DE	24.80%	15.61%
ALV.DE	20.50%	12.84%
BMW.DE	−2.39%	15.33%
BAS.DE	4.49%	14.99%
FME.DE	8.01%	16.98%

Table 5.2 Variance-covariance matrix

Covariances	DPW.DE	ALV.DE	BMW.DE	BAS.DE	FME.DE
DPW.DE	0.024376	0.011194	0.009907	0.012387	0.009991
ALV.DE	0.011194	0.016499	0.009791	0.011556	0.007837
BMW.DE	0.009907	0.009791	0.023518	0.011885	0.007562
BAS.DE	0.012387	0.011556	0.011885	0.022493	0.010644
FME.DE	0.009991	0.007837	0.007562	0.010644	0.028850

Table 5.3 Correlation matrix

Correlations	DPW.DE	ALV.DE	BMW.DE	BAS.DE	FME.DE
DPW.DE	1.000000	0.558167	0.413777	0.529003	0.376738
ALV.DE	0.558167	1.000000	0.497051	0.599888	0.359196
BMW.DE	0.413777	0.497051	1.000000	0.516735	0.290318
BAS.DE	0.529003	0.599888	0.516735	1.000000	0.417860
FME.DE	0.376738	0.359196	0.290318	0.417860	1.000000

generate (because some statistical functions give un-biased estimators while some
others give biased estimators and differ by a factor of $\frac{n}{n-1}$).

To perform a similar analysis, but this time on the 30 stocks of the DAX 30, lines
13 to 16 of Listing 5.1 must be replaced by the following lines of code of Listing 5.2

Listing 5.2 Python. Code change for 30 stocks

```
1 data = pd.Series()
2 TickerDax = ['DPW.DE', 'ALV.DE', 'BMW.DE', 'DTE.DE', 'FME.DE', 'BAS.DE
      ',
3      'HEN3.DE', 'LIN.F', 'SAP.DE', 'DBK.DE', 'BAYN.DE', 'VOW3.DE',
4      'HEI.DE', 'FRE.F', 'MRK.DE', 'BEI.DE', 'SIE.DE', 'MUV2.DE',
5      'DB1.DE', 'VNA.DE', 'EOAN.F', 'DAI.DE', 'ADS.DE', 'WDI.F',
6      'RWE.DE', 'IFX.DE', '1COV.DE', 'TKA.DE', 'CON.DE', 'LHA.DE']
7 for x in TickerDax :
8     data[x] = web.DataReader(name = x, data_source='yahoo', start=
      '2017-01-1', end='2017-12-31')
```

To add the DAX 30 to the graph, the following lines of codes can be added at the
end of the program.

Listing 5.3 Python. Adding the DAX 30

```
1 DAX = web.DataReader(name='^GDAXI', data_source='yahoo', start='
      2017-01-1', end='2017-12-31')
2 Dax_R= (DAX['Close']/DAX['Close'].shift(1)-1)
3 DaxFactor = len(Dax_R)
4 Dax_mean = np.mean(Dax_R)*DaxFactor
5 Dax_std = np.std(Dax_R)*np.sqrt(DaxFactor)
6 ax.scatter(Dax_std, Dax_mean)
7 ax.annotate("DAX", (Dax_std, Dax_mean))
```

Using now the 30 stocks of the DAX 30, we get in Fig. 5.4 the hyperbolic Frontier and the Capital Market Line. The observed return and standard deviation for the DAX 30, which can be assimilated to the Market Portfolio (as each company of the index has a weight which corresponds to its relative market capitalisation), is also added. Contrary to economic intuition, for this data set, the DAX 30 is quite different from the Tangent Portfolio. Note that here we use only the historical price returns for the assets, instead of the total returns (including the dividends), to simplify the program, but in practice people will need to use their best estimates of the future expected returns to make the allocation as accurate as possible.

5.4.2   Adding Additional Constraints

When searching for an optimal investment portfolio, some additional constraints may be imposed on the allocation π. For example, if short positions, for the risky assets, are not allowed, the condition $\forall i \in [1, d], \pi_i > 0$ will be added to the optimisation constraints. If, for diversification purposes, each stock allocation is

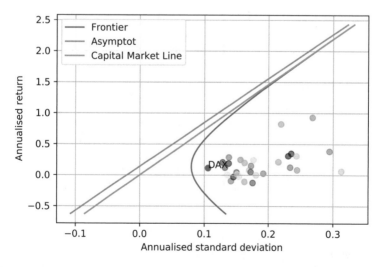

Fig. 5.4 Capital Market Line, 30 stocks

capped to 10% of the total initial wealth, then the constraint $\forall i \in [1, d]$, $\pi_i \leq 10\%$ will be added too. In these cases, usually, no explicit formula can be obtained for the optimal allocation, but numerical methods can be employed to find the solution. Also, most of these problems are quadratic optimisation problems under linear constraints, called **QP problems**, and can be solved numerically with Python.

Several Python libraries are available to solve these problems. Here, we use the "qpsolvers" library which can be run under Python 3.6. Adding this library to a computer, with Python already installed, can be done by opening CMD (Command Prompt) and typing:

< pip install qpsolvers >.

The qpsolvers library for Python solves problems of the form

$$(QPS) \begin{cases} \min_{x \in \mathbb{R}^d} \frac{1}{2}x'P\pi + q'x \\ \quad Gx \preceq h \\ \quad Ax = b, \end{cases}$$

where P is a symmetric positive definite matrix and G and A are arbitrary matrices. The notation $Gx \preceq h$ means that all the components $(Gx)_i$ of Gx and h_i of h satisfy $(Gx)_i \leq h_i$.

If we consider, for example, the problem of finding a risky investment portfolio made of the risky assets only of minimum variance, for a given expected return, with the constraints of long positions only and allocations capped to 10% for each stock, then the problem can be solved with qpsolvers. Indeed, this problem can be expressed mathematically as

$$(Q_1) \begin{cases} \min_{\pi \in \mathbb{R}^d} \frac{1}{2}\pi'\Sigma_d\pi \\ \quad M'\pi = m \\ \quad 1'_d\pi = 1 \\ \quad -\pi \preceq 0 \times 1_d \\ \quad \pi \preceq 10\% \times 1_d \end{cases}$$

and (Q_1) can be written as a (QPS) problem by taking

$$P = 2\Sigma_d \text{ and } q_1 = \begin{pmatrix} 0 \\ \vdots \\ 0 \end{pmatrix}, \quad G = \begin{pmatrix} -\text{Id}_d \\ \text{Id}_d \end{pmatrix} \text{ and } h = \begin{pmatrix} 0 \\ \vdots \\ 0 \\ 10\% \\ \vdots \\ 10\% \end{pmatrix},$$

$$A = \begin{pmatrix} m_1 & \cdots & m_d \\ 1 & \cdots & 1 \end{pmatrix} \text{ and } b = \begin{pmatrix} m \\ 1 \end{pmatrix}.$$

Table 5.4 Allocation in percentage without constraints

Stock	Alloc	Stock	Alloc	Stock	Alloc	Stock	Alloc	Stock	Alloc
DPW.DE	−7.63	HEN3.DE	1.61	HEI.DE	3.53	DB1.DE	9.02	RWE.DE	−1.21
ALV.DE	4.44	LIN.DE	7.73	FRE.F	−3.29	VNA.DE	22.19	IFX.DE	−7.05
BMW.DE	7.53	SAP.DE	2.05	MRK.DE	11.09	EOAN.F	1.96	1COV.DE	5.59
DTE.DE	15.58	DBK.DE	−3.99	BEI.DE	2.45	DAI.DE	14.59	TKA.DE	0.07
FME.DE	−1.29	BAYN.DE	2.89	SIE.DE	−7.24	ADS.DE	4.27	CON.DE	4.53
BAS.DE	−4.64	VOW3.DE	−0.65	MUV2.DE	12.93	WDI.F	3.41	LHA.DE	−0.47

Expected return 10%. Standard deviation 7.98%

Table 5.5 Allocation in percentage with positiveness constraint and 10% limit

Stock	Alloc	Stock	Alloc	Stock	Alloc	Stock	Alloc	Stock	Alloc
DPW.DE	**0**	HEN3.DE	3.18	HEI.DE	**0**	DB1.DE	9.59	RWE.DE	**0**
ALV.DE	1.93	LIN.DE	4.69	FRE.F	**0**	VNA.DE	**10.00**	IFX.DE	**0**
BMW.DE	7.78	SAP.DE	**0**	MRK.DE	**10.00**	EOAN.F	1.36	1COV.DE	4.66
DTE.DE	**10.00**	DBK.DE	**0**	BEI.DE	**10.00**	DAI.DE	**10.00**	TKA.DE	**0**
FME.DE	0.34	BAYN.DE	**0**	SIE.DE	**0**	ADS.DE	3.71	CON.DE	1.16
BAS.DE	**0**	VOW3.DE	**0**	MUV2.DE	**10.00**	WDI.F	1.62	LHA.DE	**0**

Expected return 10%. Standard deviation 8.59%

The Python code to solve (Q_1) is given in Listing 5.4 and the outputs are given in Table 5.5. The problem without constraints is also solved here, as a (QPS) problem with $G = 0$ and $h = 0$. The outputs for the unconstrained problem are given in Table 5.4 for comparison.

A few comments can be made when comparing the two allocations. First, in the unconstrained problem, some allocations are negative and some allocations are above 10%. So, in (Q_1) we expect to find some allocations reaching the 0% floor and some the 10% ceiling and show these allocations in bold in Table 5.5. In the case studied here, all the stocks with allocations above 10% in the unconstrained problem end up having allocations of 10% in the constrained problem. Secondly, most stocks having negative allocations in the unconstrained problem end up having an allocation of 0% in the constrained problem. Concerning the standard deviation, of course the unconstrained problem provides a lower limit than the constrained problem, with 7.98% instead of 8.59%.

Listing 5.4 Python. Optimal portfolio under constraints with qpsolvers

```
1  # Data extraction , calculation of the parameters Mean and Sigma
2
3  # Library importations
4  import pandas as pd
5  import pandas_datareader . data as web
6  import numpy as np
7
8  # Data extraction
```

```python
Tickers = ['DPW.DE', 'ALV.DE', 'BMW.DE', 'DTE.DE', 'FME.DE', 'BAS.DE',
        'HEN3.DE', 'LIN.F', 'SAP.DE', 'DBK.DE', 'BAYN.DE', 'VOW3.DE',
        'HEI.DE', 'FRE.F', 'MRK.DE', 'BEI.DE', 'SIE.DE', 'MUV2.DE',
        'DB1.DE', 'VNA.DE', 'EOAN.F', 'DAI.DE', 'ADS.DE', 'WDI.F',
        'RWE.DE', 'IFX.DE', '1COV.DE', 'TKA.DE', 'CON.DE', 'LHA.DE']
# 'FRE.F' changed to 'FRE.DE' after 2019/06/03
# 'WDI.F' changed to 'WDI.DE' after 2019/06/03
# 'EOAN.F' changed to 'EOAN.DE' after 2019/06/03
startinput = '2017-01-01'
endinput = '2017-12-31'
S = pd.DataFrame() # create the data frame that will contain the
    data
for t in Tickers:
S[t] = web.DataReader(name = t, data_source='yahoo', start=
    startinput, end= endinput)['Close']

# Calculation of the mean vector and variance-covariance matrix
R = pd.DataFrame()
Mean = pd.DataFrame()
Covar = pd.DataFrame()
R =S/S.shift(1)-1 # calculate the returns
R = R[1:] # eliminate the first row which is undefined
Mean = R.mean() # calculate the mean vector
Covar = R.cov() # calculate the variance-covariance matrix

# Annualisation of the mean vector and variance-covariance matrix
Mean_A = pd.DataFrame()
Covar_A = pd.DataFrame()
n = len(R) # calculate the number of returns observed
d = len(Tickers)# calculate the number of stocks used
# Calculation of the average fraction of time (expressed in years
    ) between two observations
Ys = int(startinput[0:4])
Ms = int(startinput[5:7])
Ds = int(startinput[8:10])
Ye = int(endinput[0:4])
Me = int(endinput[5:7])
De = int(endinput[8:10])
import datetime
startdate = datetime.date(Ys, Ms, Ds)
enddate = datetime.date(Ye, Me, De)
z = enddate-startdate
Duration = float(z.days) # calculate the number of calendar days
    between the two dates
m = float(len(R)-1)
delta = Duration/365/m # calculate the time interval, delta
Covar_A = Covar/delta # calculate the mean vector on an annual
    basis
Mean_A = Mean/delta # calculate the variance-covariance matrix on
    an annual basis


```

```
55 # Optimal investment portfolio of risky assets with positive
        allocations and 10% limit constraints
56 # Ref1: https://pypi.org/project/qpsolvers/
57 # Ref2: https://scaron.info/blog/quadratic-programming-in-python.
        html
58 # Ref3: https://web.stanford.edu/~boyd/papers/pdf/cvxpy_paper.pdf
59 !pip install qpsolvers # to add when running on Google Colab
60 from numpy import array, dot
61 from qpsolvers import solve_qp
62
63 m = 0.1 # expected return chosen here
64
65 P = array(2*Covar_A)
66 q = np.zeros(d)
67 G1 = np.diag([-1]*d)
68 G2 = np.diag([1]*d)
69 G = np.vstack([G1,G2])# create the matrix with blocks -Id and Id
70 h = array([0]*d+[0.1]*d)
71 A = array([Mean_A,[1]*d])
72 b = array([m,1])
73
74 alloc = solve_qp(P, q, G, h, A, b) #find the solution
75 Vec1 = [1]*d
76 print ("QP solution:", alloc)
77 print "check alloc:", dot(alloc.T, Vec1)
78 print "check mean:", dot(alloc.T, Mean_A)
79 print "standard deviation:", np.sqrt(dot(alloc.T, dot(P, alloc))
        /2) # standard deviation
```

To find the optimal portfolio with no additional constraints of positiveness or 10% limit, lines 66 to 69 of Listing 5.4 can be replaced by the two lines of Listing 5.5.

Listing 5.5 Python. Optimal portfolio without constraint with qpsolvers

```
1 G = - np.diag([0]*d)
2 h = array([0]*d)
```

5.5   Stability of the Solutions

When solving for the Markowitz optimal portfolio π^* it is possible that in the unconstrained case, some very important long and short positions may appear, making the strategy difficult to implement, and also creating a sentiment of discomfort, as it would be very aggressive to buy or sell some assets for a very large proportion of the total value of the portfolio. Here in Table 5.4 the problem is visible, even if limited, and the main issues are related to the short position of -7.63% in DPW.DE and the long position of 22.19% in VNA.DE. This being said, it is easy to understand why some very important long and short positions may appear in the solution of the unconstrained optimisation problem. Indeed, when running an optimisation with a large number of assets (taking, for example, the S&P500 for the universe of investment) or taking naturally correlated assets such as Bonds, it ends up being very likely that some of the assets will exhibit some very high correlation with others, even if sometimes incidentally because of the confidence domains of the correlation estimators. In such a situation, mathematically, long-short strategies between these extremely correlated assets will appear as offering a very high return for the risk, and above that will probably also appear as being very diversifying compared to investing in the other assets. Therefore, these long-short strategies will appear with massive weights in the optimal portfolio π^*, making the allocation cumbersome and impossible to implement. Several solutions can be considered to deal with this issue.

The first solution is to add constraints on the allocation problem, as done in Table 5.5. These constraints could also be linked to the necessity to stay close to a benchmark, in which case the fund manager may want to keep the weights of his portfolio not too far from those of the benchmark. This being said, this solution may appear as incomplete as another cause of discomfort may sometimes then arise: the number of weights in the allocation being stuck at their limits. Here in Table 5.5 out of 30 stocks, 13 are not invested in, and 6 are at their limits of 10%. To solve this issue, make things less "categorical" and allow more stocks to "have their chance", several ideas can be considered.

5.5.1 Stabilisation by Correlation Adjustment

A first practical solution that we will call **correlation adjustment** consists in correcting the correlation matrix, as we may not have enough confidence in the correlation estimates very close to 1 or minus 1, which induce massive long short strategies inside the optimal portfolio. This correlation adjustment can be obtained by multiplying all the diagonal terms of the variance-covariance matrix by $(1 + \delta^2)$, and then using this adjusted matrix for optimisation. In practice, this method should smooth the allocation results and the parameter δ can be interpreted as an extra noise added to the returns for the next period, as shown in Proposition 5.5.1.

Proposition 5.5.1 (Correlation Adjustment) *Let* $R = \mathcal{N}(M, \Sigma_d)$ *and* $\epsilon \sim \mathcal{N}(0, \mathrm{diag}(\sigma_1^2, \cdots, \sigma_d^2))$ *be independent and let* $R_\delta = R + \delta\epsilon$. *Then,*

(1) $\mathbf{Var}(R_\delta) = \mathbf{Var}(R) + \delta^2 \mathrm{diag}(\sigma_1^2, \cdots, \sigma_d^2)$,
(2) if $i \neq j$ *we get for the correlations* $\rho(R_\delta^i, R_\delta^j) = \frac{1}{1+\delta^2}\rho(R^i, R^j)$.

Proof

(1) $\mathbf{Var}(R_\delta^i) = \mathbf{Var}(R^i + \delta\epsilon_i) = \mathbf{Var}(R^i) + \delta^2\mathbf{Var}(\epsilon_i) = \sigma_i^2 + \delta^2\sigma_i^2 = (1+\delta^2)\sigma_i^2$.
   If $i \neq j$, $\mathbf{Cov}(R_\delta^i, R_\delta^j) = \mathbf{Cov}(R^i + \delta\epsilon_i, R^j + \delta\epsilon_j) = \mathbf{Cov}(R^i, R^j)$.
(2) If $i \neq j$, $\rho(R_\delta^i, R_\delta^j) = \frac{\mathbf{Cov}(R^i, R^j)}{\sqrt{(1+\delta^2)\sigma_i^2}\sqrt{(1+\delta^2)\sigma_j^2}} = \frac{1}{1+\delta^2}\rho(R^i, R^j)$. $\qquad\square$

Table 5.6 Allocation in percentage with positiveness, 10% limit and 10% shrinkage

Stock	Alloc	Stock	Alloc	Stock	Alloc	Stock	Alloc	Stock	Alloc
DPW.DE	**0**	HEN3.DE	4.16	HEI.DE	0.29	DB1.DE	9.59	RWE.DE	**0**
ALV.DE	4.06	LIN.DE	4.49	FRE.F	**0**	VNA.DE	**10.00**	IFX.DE	**0**
BMW.DE	6.18	SAP.DE	1.86	MRK.DE	**10.00**	EOAN.F	1.44	1COV.DE	3.92
DTE.DE	**10.00**	DBK.DE	**0**	BEI.DE	8.27	DAI.DE	**10.00**	TKA.DE	**0**
FME.DE	1.18	BAYN.DE	0.05	SIE.DE	**0**	ADS.DE	3.34	CON.DE	2.07
BAS.DE	**0**	VOW3.DE	0.38	MUV2.DE	8.54	WDI.F	1.57	LHA.DE	**0**

Expected return 10%. Standard deviation 8.81%

Table 5.6 shows the effect of an adjustment by a factor $\delta = 10\%$ for allocations limited to the range [0%, 10%]. Compared to Table 5.5, the number of allocations at 0% is reduced from 13 to 9, and the number of allocation at 0% is reduced from 6 to 4. These allocations with limit values 0 and 10 appear in bold in Table 5.6. The optimal portfolio sees its standard deviation increasing from 8.59 to 8.81% as we have added some noise to the model.

5.6 The Bayesian Approach

In practice, another way to look at the whole process of finding optimal portfolios is through **Bayesian statistics**, which takes into account the estimation risk for the model parameters M and Σ. So far, the search for an optimal portfolio has been performed in two steps.

- Step 1, solving the optimisation problem assuming that M and Σ are known.
- Step 2, replacing the parameters in the solution of Step 1 by their estimates.

However, this two-step approach may be suboptimal, because the problems of optimisation and estimation are not separate, but intricate, and therefore, ideally, should be solved simultaneously. For example, if π^* is the best portfolio for the parameters $\widehat{M}$ and $\widehat{\Sigma}$ and if π^{**} performs better than π^* for all the other possible values of M and Σ then, allocating according to π^{**}, rather than π^*, may seem wiser and more robust, because of the remaining uncertainty around the true values of M and Σ after estimation.

To define the best portfolios under model uncertainty we come back to the utility function paradigm and start by establishing in Proposition 5.6.1, in the Markowitz context, the link between maximising a utility function and finding an optimal portfolio.

Proposition 5.6.1 (Utility and Markowitz Mean-Variance Optimisation) *In an economy with a risk-free asset r_0, where $R(\pi)$ denotes the return of an investment portfolio of risky allocation π, the problem defined for $\lambda > 0$ by*

$$(U_\lambda) : \max_{\pi \in \mathbb{R}^d} \mathbf{E}(R(\pi)) - \frac{\lambda}{2} \mathbf{Var}(R(\pi))$$

has a unique solution π_λ and this solution corresponds to the risky allocation of a Markowitz optimal portfolio, of expected return

$$\mathbf{E}(R(\pi_\lambda)) = r_0 + \frac{1}{\lambda} \|M - r_0 1_d\|^2_{\Sigma_d^{-1}}.$$

Proof If π is the risky allocation of an investment portfolio then $1 - 1_d'\pi$ is its risk-free allocation and we get

$$R(\pi) = (1 - 1_d'\pi)r_0 + \pi' R = r_0 + \pi'(R - r_0 1_d)$$

and from there

$$\mathbf{E}(R(\pi)) = r_0 + \pi'(M - r_0 1_d)$$

and

$$\mathbf{Var}(R(\pi)) = \pi' \Sigma_d \pi.$$

So, if we define

$$U_\lambda(\pi) = \mathbf{E}(R(\pi)) - \frac{\lambda}{2} \mathbf{Var}(R(\pi))$$

we get

$$\frac{\partial U_\lambda}{\partial \pi} = 0 \Longleftrightarrow (M - r_0 1_d) - \lambda \Sigma \pi = 0$$

$$\Longleftrightarrow \pi = \frac{1}{\lambda} \Sigma^{-1}(M - r_0 1_d),$$

which corresponds to the risky allocation of a Markowitz optimal portfolio according to Theorem 5.2.1. Moreover, this portfolio has a return of

$$M'\pi + (1 - 1_d'\pi)r_0 = r_0 + (M - r_0 1_d)'\pi = r_0 + \frac{1}{\lambda}\|M - r_0 1_d\|^2_{\Sigma^{-1}},$$

which finishes the proof.   □

From now on, when maximising $U_\lambda(\pi)$, we want to take into account the fact that the parameters M and Σ are unknown. For that, mathematically, we place ourselves in a **Bayesian framework** where:

- the law of R, conditionally on M and Σ, is a Gaussian law $\mathcal{N}(M, \Sigma)$,
- there is an a priori probability μ_0 on (M, Σ), before any observation is made,
- the information available after observing the n vector of returns $(R_i)_{i \in [\![1,n]\!]}$ over the period $[\![0, n]\!]$, or after having gathered any other type of information during this period, is denoted $\mathcal{F}_n$.

If R_{n+1} denotes the vector of returns for the assets over the next period, the problem (U_λ) of maximising the utility now becomes in its Bayesian version

$$(U_\lambda^B) : \max_{\pi \in \mathbb{R}^d} \mathbf{E}(R_{n+1}(\pi)|\mathcal{F}_n) - \frac{\lambda}{2}\mathbf{Var}(R_{n+1}(\pi)|\mathcal{F}_n) \qquad (5.6.1)$$

and the solution of (U_λ^B) is given in Property 5.6.1.

Property 5.6.1 (Utility and Bayesian Mean-Variance Optimisation) In a Bayesian framework, where M and Σ are unknown (and r_0 is known), and where $R(\pi)$ represents the return of an investment portfolio of risky allocation π, then the problem

$$(U_\lambda^B) : \max_{\pi \in \mathbb{R}^d} \mathbf{E}(R_{n+1}(\pi)|\mathcal{F}_n) - \frac{\lambda}{2}\mathbf{Var}(R_{n+1}(\pi)|\mathcal{F}_n)$$

has a unique solution π_λ^B defined by

$$\pi_\lambda^B = \frac{1}{\lambda}\mathbf{Var}(R_{n+1}|\mathcal{F}_n)^{-1}\mathbf{E}(R_{n+1} - r_0 1_d|\mathcal{F}_n). \qquad (5.6.2)$$

Proof As in Proposition 5.6.1 we get

$$R_{n+1}(\pi) = (1 - 1_d'\pi)r_0 + \pi'R_{n+1} = r_0 + \pi'(R_{n+1} - r_0 1_d)$$

and from there

$$\mathbf{E}(R_{n+1}(\pi)|\mathcal{F}_n) = r_0 + \pi'\mathbf{E}(R_{n+1} - r_0 1_d|\mathcal{F}_n)$$

and

$$\mathbf{Var}(R_{n+1}(\pi)|\mathcal{F}_n) = \pi'\mathbf{Var}(R_{n+1}|\mathcal{F}_n)\pi.$$

Therefore, if we note

$$U_\lambda^B(\pi) = r_0 + \pi' \mathbf{E}(R_{n+1} - r_0 1_d | \mathcal{F}_n) - \frac{\lambda}{2}\pi' \mathbf{Var}(R_{n+1}|\mathcal{F}_n)$$

to maximise $U_\lambda^B(\pi)$ we solve

$$\frac{\partial U_\lambda^B}{\partial \pi} = 0$$

$$\Longleftrightarrow \mathbf{E}(R_{n+1} - r_0 1_d | \mathcal{F}_n) - \lambda \mathbf{Var}(R_{n+1}|\mathcal{F}_n)\pi = 0$$

$$\Longleftrightarrow \pi = \frac{1}{\lambda}\mathbf{Var}(R_{n+1}|\mathcal{F}_n)^{-1}\mathbf{E}(R_{n+1} - r_0 1_d|\mathcal{F}_n), \qquad (5.6.3)$$

which proves the result.   □

So, to find the **Bayesian optimal portfolio** solutions of (U_λ^B), the issue is to calculate $\mathbf{Var}(R_{n+1}|\mathcal{F}_n)$ and $\mathbf{E}(R_{n+1}|\mathcal{F}_n)$. Note that, in the Bayesian models discussed below, either both M and Σ are assumed to be unknown parameters or M alone, in which case the prior μ_0 is on M alone and the result from Eq. (5.6.3) subsists. When a parameter is unknown, its estimate will come, in the cases discussed below, either from its empirical estimate or from a market consensus, twisted by the a priori law. The resulting calculations can be complex, depending on the choice of the prior distribution μ_0. In the next sections we present some results for three types of priors. The **Jeffrey's prior** on (M, Σ), a Gaussian prior on M (with Σ assumed to be known) and a prior on M derived from the **Black–Litterman** model, in which Σ is assumed to be known.

5.6.1 Jeffrey's Prior μ_0 on M and Σ

Several (prior) distributions for (M, Σ) can be considered. Amongst these distributions, the **conjugate prior** and **Jorion's prior** are discussed in Avramov and Zhou [9]. Here we discuss **Jeffrey's prior**, which is considered as the most noninformative prior on (M, Σ) (see Sabanes Bove [79] for more on the topic). The proofs of the results presented here can be found in Bauder et al. [17]. Here, d represents the number of risky assets, and n represents the number of observed vectors of returns.

Let $\widehat{M}$ and $\widehat{\Sigma}$ be the empirical estimators for M and Σ, as defined in Theorem 3.3.2, based on the observations of the $(R_i)_{i \in [\![1,n]\!]}$, which represents the information $\mathcal{F}_n$. Then, for Jeffrey's prior the result are as follows:

- $\mathbf{E}(R_{n+1}|\mathcal{F}_n) = \widehat{M}$,
- $\mathbf{Var}(R_{n+1}|\mathcal{F}_n) = c_{d,n} \times \widehat{\Sigma}$ with $c_{d,n} = \frac{n}{n-d-1} + \frac{2n-d-1}{n(n-d-1)(n-d-2)}$.

So, with Jeffrey's prior, the empirical estimate of M is not twisted and the Bayesian approach leads to results similar in nature to the ones obtained with the two-step optimisation process. The only adjustment to be made is for the variance-covariance matrix estimate, which has to be multiplied by the constant factor $c_{d,n}$: As a consequence, the correlation matrix used is unaffected and only the individual standard deviation estimates $\widehat{\sigma}_i$ have to be multiplied by the constant factor $\sqrt{c_{d,n}}$.

Other than that, according to Corollary 3.3.2

$$\mathbf{E}\left(\frac{n-d-2}{n}\widehat{\Sigma}_d^{-1}\right) = \Sigma_d^{-1},$$

so a few remarks can be made about the correction factor $c_{d,n}$:

- the first term $\frac{n}{n-d-1}$ partially corrects the bias of $\widehat{\Sigma}_d^{-1}$,
- the second term $\frac{2n-d-1}{n(n-d-1)(n-d-2)}$ is of second order,
- the perception of the risk is higher in the Bayesian optimisation as $c_{d,n} > 1$ but, as n tends to infinity, $c_{d,n}$ converges to 1 and the Bayesian results converge towards those of the standard (two-step) approach.

So, the main takeaway is that an estimator of Σ_d^{-1} corrected partially for the bias should be used in the standard two-step procedure. Note also that, as $\frac{d}{n}$ approaches 1, the bias for $\widehat{\Sigma}_d^{-1}$ is large and both Bayesian and non-Bayesian statisticians should be concerned with the quality of the estimators (biased or de-biased) for Σ^{-1}, as the number of observations may be considered as small compared to the number of parameters to estimate.

Finally, we make the following observations:

- the frontiers $\mathcal{F}$, for the risky investment portfolios, in the risk-return representations, based on the mean vector and variance-covariance matrix estimates, coincide for the Bayesian and standard approach, after rescaling the risk axis by the factor $\sqrt{c_{d,n}}$.
- The Tangent Portfolio in the standard approach has weights (summing to 1) proportional to $\widehat{\Sigma}_d^{-1}(\widehat{M} - r_0 1_d)$ and in the Bayesian approach proportional to $c_{d,n}\widehat{\Sigma}_d^{-1}(\widehat{M} - r_0 1_d)$, consequently it has the same allocation in both frameworks.
- The Tangent Portfolio has an estimated risk of $\sqrt{\pi^{*\prime}\widehat{\Sigma}\pi^*}$ in the standard approach and of $\sqrt{\pi^{*\prime}c_{d,n}\widehat{\Sigma}\pi^*}$ in the Bayesian approach, so without surprise the risk is multiplied by the factor $\sqrt{c_{d,n}}$.
- The maximum Sharpe ratio is reached for the Tangent Portfolio and therefore deteriorates by the factor $\sqrt{c_{d,n}}$ in the (more conservative) Bayesian approach.

Numerical Examples

For the DAX 30, with 253 returns observed over 1 year for 30 stocks, the Bayesian adjustment factor turns out to be $c_{30,253} = 1.149$, which means that each individual empirical standard deviation estimate $\widehat{\sigma}_i$ has to be multiplied by the factor $\sqrt{c_{30,253}} = 1.072$ for the estimation risk.

For the Japanese index Nikkei 225 with 253 daily returns observed over a year for 225 components we get $\sqrt{c_{225,253}} = 3.125$. So, from a Bayesian perspective, the standard two-step optimisation seems problematic, as it seems to overestimate the maximum achievable Sharpe Ratio by the factor 3.125.

So, when n is not significantly larger than d and $c_{d,n}$ deviates significantly from 1, it seems safer to run an optimisation on a simplified model, where some assets are grouped together, for example in sub-indices which could be industry related (and which could be optimised as well) or factor related. This would lead to a top-down optimisation approach which would certainly be more robust and provide similar results for the standard and the Bayesian approach.

5.6.2   Gaussian Prior μ_0 on M

Here we study a Bayesian model in which the variance-covariance matrix Σ is assumed to be a known parameter and the vector of expected returns M is an unknown parameter coming from a normal distribution. The model can be written as

$$(BE) \begin{cases} \mathcal{L}(R|M) \sim \mathcal{N}(M, \Sigma) \text{ with } \Sigma \text{ known and} \\ \mathcal{L}(M) \sim \mathcal{N}(\mu, \Omega) \text{ with } \mu \text{ and } \Omega \text{ known.} \end{cases}$$

$\mathcal{L}(R|M)$ denotes the law of the vector of returns, conditionally on the knowledge of M, and $\mathcal{L}(M)$ is the (prior) law of M. We denote by $(R_i)_{i\in[\![1,n]\!]}$ the vectors of returns, for the assets, observed over n periods and

$$R_{[n]} = (R_1, \cdots, R_n).$$

All the R_i are assumed to depend on the same realisation of M and to be independent conditionally on M with the law of $\mathcal{L}(R|M)$. As seen in Eq. (5.6.2) the Bayesian optimal portfolio π_λ for the period $n + 1$ is defined by

$$\pi_\lambda = \frac{1}{\lambda} \mathbf{Var}(R_{n+1}|R_{[n]} = r_{[n]})^{-1} \mathbf{E}(R_{n+1} - r_0 1_d | R_{[n]} = r_{[n]}).$$

To calculate this quantity we use Proposition 5.6.2:

Proposition 5.6.2 (Bayesian Optimal Portfolio) *In the (BE) model the following properties are satisfied:*

(1) $\mathbf{E}(R_{n+1}|R_{[n]} = r_{[n]}) = \mathbf{E}(M|R_{[n]} = r_{[n]})$.
(2) $\mathbf{Var}(R_{n+1}|R_{[n]} = r_{[n]}) = \mathbf{Var}(M|R_{[n]} = r_{[n]}) + \Sigma$.
(3) $(M, R_1, \cdots, R_n)$ *is a Gaussian vector.*
(4) $\mathcal{L}(M|R_1, R_2, \cdots R_n)$ *is a Gaussian law.*

(5) $\mathbf{E}(M|R_{[n]} = r_{[n]}) = \left(\Sigma^{-1} + \frac{\Omega^{-1}}{n} \right)^{-1} \left(\Sigma^{-1} \bar{r} + \frac{\Omega^{-1}}{n} \mu \right)$ with $\bar{r} = \frac{1}{n} \sum_{i=1}^{i=n} r_i$.

(6) $\mathbf{Var}(M|R_{[n]} = r_{[n]}) = \frac{1}{n} (\Sigma^{-1} + \frac{\Omega^{-1}}{n})^{-1}$.

Proof For (1) and (2) we write $R_{n+1} = M + Z$ with $Z \sim \mathcal{N}(0, \Sigma)$ independent from all the other variables. Therefore, by linearity of the expectation and independence,

$$
\begin{aligned}
\mathbf{E}(R_{n+1}|R_{[n]} = r_{[n]}) &= \mathbf{E}(M + Z|R_{[n]} = r_{[n]}) \\
&= \mathbf{E}(M|R_{[n]} = r_{[n]}) + \mathbf{E}(Z|R_{[n]} = r_{[n]}) \\
&= \mathbf{E}(M|R_{[n]} = r_{[n]}) + \mathbf{E}(Z) \\
&= \mathbf{E}(M|R_{[n]} = r_{[n]})
\end{aligned}
$$

and by additivity of the variance for independent variables,

$$
\begin{aligned}
\mathbf{Var}(R_{n+1}|R_{[n]} = r_{[n]}) &= \mathbf{Var}(M + Z|R_{[n]} = r_{[n]}) \\
&= \mathbf{Var}(M|R_{[n]} = r_{[n]}) + \mathbf{Var}(Z|R_{[n]} = r_{[n]}) \\
&= \mathbf{Var}(M|R_{[n]} = r_{[n]}) + \Sigma.
\end{aligned}
$$

For (3) we write $R_i = M + Z_i$ with $Z_i \sim \mathcal{N}(0, \Sigma)$ independent from all the other variables. $(M, Z_1 \cdots, Z_n)$ is Gaussian and as $(M, R_1 \cdots, R_n)$ is a linear transformation of $(M, Z_1 \cdots, Z_n)$ it is Gaussian as well, which proves the result.

For (4), (5), (6), according to Bayes' rules,

$$
P(M = m|R_{[n]} = r_{[n]}) \propto P(R_{[n]} = r_{[n]}|M = m) P(M = m),
$$

$$
\propto \exp \left(- \sum_{i=1}^{n} (r_i - m)' \frac{\Sigma^{-1}}{2} (r_i - m) \right) \exp \left(- (m - \mu)' \frac{\Omega^{-1}}{2} (m - \mu) \right),
$$

$$
\propto \exp \left(- \sum_{i=1}^{n} (m - r_i)' \frac{\Sigma^{-1}}{2} (m - r_i) \right) \exp \left(- (m - \mu)' \frac{\Omega^{-1}}{2} (m - \mu) \right).
$$

Now, if we try to factorise this expression in m, up to a proportionality constant we find

$$
\propto \exp \left(- (m - a_n)' \frac{n \Sigma^{-1} + \Omega^{-1}}{2} (m - a_n) \right), \tag{5.6.4}
$$

where a_n is defined by

$$(n\Sigma^{-1} + \Omega^{-1})a_n = \Sigma^{-1}(\sum_{i=1}^{i=n} r_i) + \Omega^{-1}\mu$$

$$\Longrightarrow a_n = \left(\Sigma^{-1} + \frac{\Omega^{-1}}{n}\right)^{-1}\left(\Sigma^{-1}\bar{r} + \frac{\Omega^{-1}\mu}{n}\right)$$

with $\bar{r} = \frac{1}{n}\sum_{i=1}^{i=n} r_i$.

Equation (5.6.4) shows that $\mathcal{L}(M|R_1, \cdots R_n)$ is a Gaussian law and from this density expression (up to a constant multiplication factor) we can deduce its expectation, which is a_n, and its variance-covariance matrix, which is $(n\Sigma^{-1} + \Omega^{-1})^{-1}$, which finishes the proof.   □

Corollary 5.6.1

$$\mathbf{E}(M|R_{[n]} = r_{[n]}) = \bar{r} + \frac{1}{n}\left(\Sigma^{-1} + \frac{\Omega^{-1}}{n}\right)^{-1}\Omega^{-1}(\mu - \bar{r}).$$

Proof

$$\mathbf{E}(M|R_{[n]} = r_{[n]}) - \bar{r}$$

$$= \mathbf{E}(M|R_{[n]} = r_{[n]}) - \left(\Sigma^{-1} + \frac{\Omega^{-1}}{n}\right)^{-1}\left(\Sigma^{-1} + \frac{\Omega^{-1}}{n}\right)\bar{r}$$

$$= \left(\Sigma^{-1} + \frac{\Omega^{-1}}{n}\right)^{-1}\left[\Sigma^{-1}\bar{r} + \frac{\Omega^{-1}\mu}{n} - \Sigma^{-1}\bar{r} - \frac{\Omega^{-1}}{n}\bar{r}\right]$$

$$= \left(\Sigma^{-1} + \frac{\Omega^{-1}}{n}\right)^{-1}\frac{\Omega^{-1}}{n}(\mu - \bar{r}).$$ □

Note that when n is large, we can deduce from Property 5.6.2 that

- $\text{Var}(R_{n+1}|R_{[n]} = r_{[n]})$ is close to Σ (which is assumed to be known),
- $\mathbf{E}(R_{n+1}|R_{[n]} = r_{[n]})$ is close to $\bar{r}$ and
- π_λ is close to $\frac{1}{\lambda}\Sigma^{-1}(\bar{r} - r_0 1_d)$.

Therefore, if there is a large number of observations, the outcomes of this theoretical Bayesian model are similar to the standard Markowitz model, with for both models the problem of assuming that Σ is known (while it is not in reality) and the replacement of M by its empirical estimate $\bar{r}$ when calculating the optimal portfolios in practice.

5.6.3 The Black–Litterman Model

The **Black–Litterman** model [21] is equivalent to a Bayesian model on M using a prior distribution linked, after a linear transformation, to a Gaussian law. Here the estimates are not updated through the observations of some past returns but via a market consensus C. The assumptions are that:

$$(BL) \begin{cases} C \text{ and } R \text{ are independent conditionally on } M \\ \mathcal{L}(R|M) \sim \mathcal{N}(M, \Sigma), \text{ with } \Sigma \text{ known} \\ \mathcal{L}(C|M) \sim \mathcal{N}(M, \Sigma_\tau), \text{ with } \Sigma_\tau \text{ known} \\ P(M = m) \propto P(V = Qm) \text{ with } Q \text{ known and} \\ \mathcal{L}(V) \sim \mathcal{N}(Q\mu, \Omega) \text{ with } \mu \text{ and } \Omega \text{ known invertible,} \end{cases}$$

where Q is a $k \times d$-matrix of rank k (full row rank). So, this model is equivalent to a Bayesian model on M where the prior $\mu_0(\cdot)$ for M is defined by $\mu_0(m) \propto P(V = Qm)$. Note that, it is not restrictive to write the expected value of V in the form $Q\mu$ as Q is full row rank and we use this format because it simplifies the calculations. As before, the risky allocation for an optimal portfolio, when taking into account the market consensus C, is of the form

$$\pi_\lambda = \frac{1}{\lambda} \mathbf{Var}(R|C = c)^{-1} \mathbf{E}(R - r_0 1_d | C = c)$$

and the issue is to calculate $\mathbf{E}(R|C = c)$ and $\mathbf{Var}(R|C = c)$.

For this we use the results of Proposition 5.6.3.

Proposition 5.6.3 (Black–Litterman Optimal Portfolio) *Let* $\Omega_Q^{-1} = Q'\Omega^{-1}Q$ *in the model. Then we have:*

(1) $\mathbf{E}(R|C = c) = \mathbf{E}(M|C = c)$,
(2) $\mathbf{Var}(R|C = c) = \mathbf{Var}(M|C = c) + \Sigma$,
(3) $\mathbf{E}(M|C = c) = \left(\Sigma_\tau^{-1} + \Omega_Q^{-1} \right)^{-1} \left(\Sigma_\tau^{-1} c + \Omega_Q^{-1} \mu \right)$,
(4) $\mathbf{Var}(M|C = c) = (\Sigma_\tau^{-1} + \Omega_Q^{-1})^{-1}$.

Proof The proofs are similar to those of Property 5.6.2.
For (1) and (2) we write $R = M + Z$ where $Z \sim \mathcal{N}(0, \Sigma)$ and Z is independent from M and C. By linearity of the expectation and independence of Z and C,

$$\mathbf{E}(R|C = c) = \mathbf{E}(M|C = c) + \mathbf{E}(Z|C = c)$$
$$= \mathbf{E}(M|C = c) + \mathbf{E}(Z) = \mathbf{E}(M|C = c),$$

and by independence of the variables,

$$\mathbf{Var}(R|C = c) = \mathbf{Var}(M + Z|C = c)$$
$$= \mathbf{Var}(M|C = c) + \mathbf{Var}(Z|C = c)$$
$$= \mathbf{Var}(M|C = c) + \Sigma.$$

For (3) and (4), according to Bayes' rules we have

$$P(M = m|C = c) \propto P(C = c|M = m)P(V = Qm)$$

$$\propto \exp\left(-(c - m)'\frac{\Sigma_\tau^{-1}}{2}(c - m)\right)\exp\left(-(Qm - Q\mu)'\frac{\Omega^{-1}}{2}(Qm - Q\mu)\right)$$

$$\propto \exp\left(-(m - c)'\frac{\Sigma_\tau^{-1}}{2}(m - c)\right)\exp\left(-(m - \mu)'\frac{\Omega_Q^{-1}}{2}(m - \mu)\right)$$

which factorises as in Proposition 5.6.2 as the density of a normal distribution

$$\propto \exp\left(-(m - b_n)'\frac{(\Sigma_\tau^{-1} + \Omega_Q^{-1})}{2}(m - b_n)\right), \qquad (5.6.5)$$

where b_n, which represents the mean, is defined by

$$(\Sigma_\tau^{-1} + \Omega_Q^{-1})b_n = \Sigma_\tau^{-1}c + \Omega_Q^{-1}\mu$$

$$\implies b_n = (\Sigma_\tau^{-1} + \Omega_Q^{-1})^{-1}\left(\Sigma_\tau^{-1}c + \Omega_Q^{-1}\mu\right)$$

and the matrix of variance-covariance for this Gaussian distribution appears from Eq. (5.6.4) to be $(\Sigma_\tau^{-1} + \Omega_Q^{-1})^{-1}$, which finishes the proof. □

Corollary 5.6.2

$$\mathbf{E}(M|C = c) = c + \left(\Sigma_\tau^{-1} + \Omega_Q^{-1}\right)^{-1}\Omega_Q^{-1}(\eta - c).$$

Proof The method is the same as in Corollary 5.6.1.

$$\mathbf{E}(M|C = c) - c = \mathbf{E}(M|C = c) - \left(\Sigma_\tau^{-1} + \Omega_Q^{-1}\right)^{-1}\left(\Sigma_\tau^{-1} + \Omega_Q^{-1}\right)c$$

$$= \left(\Sigma_\tau^{-1} + \Omega_Q^{-1}\right)^{-1}\left(\left(\Sigma_\tau^{-1}c + \Omega_Q^{-1}\mu\right) - \left(\Sigma_\tau^{-1} + \Omega_Q^{-1}\right)c\right)$$

$$= \left(\Sigma_\tau^{-1} + \Omega_Q^{-1}\right)^{-1}\Omega_Q^{-1}(\mu - c). \qquad □$$

Note that the matrix Q represents the fund manager's personal views on some stocks, in relative or in absolute terms, and these views are modelised by a Gaussian distribution $\mathcal{N}(Q\mu, \Omega)$ attributed to the future performances of some particular investment or self-financing portfolios.

For example:

- If the first view is that stock 1 has an expected return 2% above stock 2 then in the first row of Q the first component will be 1 and the second component -1, and the first component of $Q\mu$ will be 2%.
- If the second view is that stock 2 has an expected return of 5% then in the second row of Q the second coefficient will be 1 (and the others 0) and the second component of $Q\mu$ will be 5%.

There is also a certain degree of confidence attributed by the fund manager to his views and this translates into the choice of a matrix Ω.

If the asset manager has no specific views then the term Ω_Q^{-1} can be replaced by zero in Proposition 5.6.3, leading to

(3) $\mathbf{E}(R|C = c) = c,$
(4) $\mathbf{Var}(R|C = c) = \Sigma_\tau + \Sigma,$
(5) $\pi_\lambda = \frac{1}{\lambda}(\Sigma + \Sigma_\tau)^{-1}(c - r_0 1_d).$

In this case, the **Market Portfolio** of risky allocation π_M is the **Tangent Portfolio** if and only if, according to Proposition 5.2.2,

$$c - r_0 1_d \propto (\Sigma_\tau + \Sigma)\pi_M$$

and such a value of c is denoted c_e and called an **equilibrium return**.

So, in the Black–Litterman model if the base assumption is that the market consensus C is an equilibrium value c_e, then it is only through the specific views V that deviations from the **Market Portfolio** will be obtained.

A Few References

1. Avramov, D., & Zhou, G., (2010). Bayesian portfolio analysis. *The Annual Review of Financial Economics, 2*, 25–47.
2. Bauder, D., Bodnar, T., Parolya, N., & Schmid, W. (2018). Bayesian mean-variance analysis: Optimal portfolio selection under parameter uncertainty. arXiv:1803.03573v1 [q-fin.ST]. 9 March 2018.
3. Black, F. (1972). Capital market equilibrium with restricted borrowing. *The Journal of Business,45*(3), 445–455.
4. Black, F., & Litterman, R. (1992). Global optimization. *Financial Analysts Journal, 48*(5), Sep/Oct 1992.

5. Fama, E. F., & French, K. R. (2004). The capital asset pricing model: Theory and evidence. *Journal of Economic Perspectives, 18*(3), 25–46.
6. Lintner, J. (1965). The valuation of risk assets and the selection of risky investments in stock portfolios and capital budgets. *The Review of Economics and Statistics, 47*, 13–37.
7. Mossin, J. (1966, October). Equilibrium in a capital asset market. *Econometrica, 34*(4), 768–783.
8. Roll, R. (1977). A critique of the asset pricing theory's tests' part I: On past and potential testability of the theory. *Journal of Financial Economics, 4*(2), 129–176.
9. Sabanes Bove, D., & Held, L. (2014). *Applied statistical inference.* Heidelberg: Springer.
10. Sharpe, W. F. (1964). Capital asset prices: A theory of market equilibrium under conditions of risk. *The Journal of Finance, 19*(3), 425–442.
11. Tobin, J. (1958, February). Liquidity preference as behavior towards risk. *Review of Economic Studies, XXV*(2), 65–86, HB1R4.

Performance and Diversification Indicators

6

This chapter describes some statistics which, when screening a large number of funds, are useful to classify them and to automatically select the ones which seem to be particularly relevant. Some of these indicators depend on the leverage used by the funds while others measure the intrinsic quality of the fund, i.e. its engine of performance independently from any potential leverage artefacts. The **Diversification ratio** is also explained, as it is linked to many new alternative methods of asset allocation such as **risk parity** investing.

6.1   The Sharpe Ratio

Definition 6.1.1 (Sharpe Ratio) The **Sharpe ratio** of an investment portfolio P is defined as $\frac{m_P - r_0}{\sigma_P}$.

Remark 6.1.1 Under the Markowitz's framework:

- The Sharpe ratio is maximal for portfolios belonging to the Capital Market Line.
- All the portfolios belonging to the Capital Market Line have the same Sharpe ratio, which is equal to the slope of the Capital Market Line.
- When investing in a single portfolio, wealth should be allocated first by determining a portfolio with the maximum Sharpe ratio and then by allocating all the wealth between this portfolio and the risk free asset.

Proposition 6.1.1 *The Sharpe ratio is independent of the leverage as the portfolio* $\lambda \Pi_P + (1 - \lambda)\Pi_0$ *has the same Sharpe ratio as the portfolio* Π_P *for any* $\lambda > 0$.

Proof Let Π_λ denote the portfolio $\lambda \Pi_P + (1 - \lambda)\Pi_0$. We have $\mathbf{E}(R(\Pi_\lambda)) = \mathbf{E}(\lambda R(\Pi_P) + (1 - \lambda)r_0)$ so, $\mathbf{E}(R(\Pi_\lambda)) - r_0 = \lambda \mathbf{E}(R(\Pi_P) - r_0)$ and $\mathbf{Var}(R(\Pi_\lambda)) = \mathbf{Var}(\lambda R(\Pi_P)) = \lambda^2 \mathbf{Var}(R(\Pi_P))$. So, $\sigma(R(\Pi_\lambda)) = \lambda \sigma(R(\Pi_P))$ for $\lambda > 0$ and the

© Springer Nature Switzerland AG 2020
P. Brugière, *Quantitative Portfolio Management*, Springer Texts in Business and Economics, https://doi.org/10.1007/978-3-030-37740-3_6

Sharpe ratio for Π_λ is

$$\frac{\mathbf{E}(R(\Pi_\lambda)) - r_0}{\sigma(R(\Pi_\lambda))} = \frac{\lambda \mathbf{E}(R(\Pi_P)) - r_0)}{\lambda \sigma(R(\Pi_P))} = \frac{\mathbf{E}(R(\Pi_P)) - r_0)}{\sigma(R(\Pi_P))}. \qquad \square$$

Remark 6.1.2 Some asset managers sometimes offer to investors different versions of the same fund with different leverages, as it is not always easy for an investor to leverage himself (i.e. to borrow to take more exposure to a fund). In this case on a risk/return representation these funds clearly appear on the same line.

Remark 6.1.3 The Sharpe ratio is usually estimated by $\frac{\hat{m}_P - r_0}{\hat{\sigma}_P}$ where $\hat{m}_P$ and $\hat{\sigma}_P$ are historical estimates of m and σ.

6.2   The Jensen Index

Definition 6.2.1 (Jensen Index) The **Jensen Index** of an investment portfolio P is defined as $m_P - \left(r_0 + \beta_T(P)(m_T - r_0)\right)$.

Remark 6.2.1 Under the Markowitz's framework this quantity should be zero according to the Security Market Line. In practice, this quantity is called the alpha of the fund.

Remark 6.2.2 In practice:

- the betas are estimated historically,
- the expected returns are either historical estimates or analyst predictions, usually based on a target price for the stock at a horizon of a few months.

Remark 6.2.3 ("Pair Trading") In practice, a trader may consider the following pair-trading strategy:

- represent all the stocks he can buy or sell (his trading universe) by their (β, m) estimates. The beta will be calculated with respect to an index to which all these stocks belong,
- determine a regression line for the estimated (β, m), which in practice will not be perfectly aligned, even if in the model the (unknown) parameters (β, m) should be aligned,
- build a long/short portfolio where the stocks above the SML are purchased and the stocks below the SML are sold.

In this strategy, the trader expects that the stocks away from the regression line are going to return there. As the target prices are fixed, it is via a change in the spot prices that the return to the regression line is expected to happen. Therefore, securities above the regression line (which are the ones purchased) are expected to appreciate (for their expected returns to reduce) while the ones below the regression line are expected to depreciate. This type of strategy is called "long/short" or "relative value" or "pair trading" when it is put in place only for a pair of stocks. An interesting characteristic of these long/short strategies is their low betas to the market, which should ultimately make it possible to make some gains, independently of the evolution of the broader index itself.

Remark 6.2.4 A large pension fund which allocates money amongst many asset managers may assume that the idiosyncratic risk is going to be reduced/cancelled through diversification and in this case may be concerned only by the remuneration of the non-diversifiable risk and by the Jensen Index of each fund. Here, the excess return (alpha) is usually not seen as a market anomaly but as something which is expected to last and which is linked to some particular skills from the asset manager.

Property 6.2.1 The Jensen Index depends on the leverage.

Proof Let P_λ be the portfolio of allocation $\Pi_\lambda = \lambda \Pi_P + (1 - \lambda)\Pi_0$. The Jensen Index of P_λ is $m_{P_\lambda} - (r_0 + \beta_T(P_\lambda)(m_T - r_0))$. As $m_{P_\lambda} = \lambda m_P + (1 - \lambda)r_0$ and $\beta_T(P_\lambda) = \frac{\text{Cov}(R_{P_\lambda}, R_T)}{\sigma^2(R_T)} = \lambda \frac{\text{Cov}(R_P, R_T)}{\sigma^2(R_T)} = \lambda \beta_T(P)$, we get

$$\lambda m_P - \lambda r_0 - \lambda \beta_T(P_\lambda)(m_T - r_0) = \lambda\big(m_P - r_0 - \beta_T(P_\lambda)(m_T - r_0)\big). \qquad \square$$

Exercise 6.2.1 How do you read in a beta/return representation, $\{(\beta, m)\}$, the Jensen Index of a fund?

6.3 The Treynor Index

Definition 6.3.1 (Treynor Index) The **Treynor Index** of an investment portfolio P is defined as $\frac{m_P - r_0}{\beta_T(P)}$.

Remark 6.3.1 Under the Markowitz framework the Treynor Index should be constant according to the SML. The Treynor Index is similar to the Jensen Index in its objective to detect funds for which there is an excess of remuneration of the systematic risk. Compared to the Jensen Index the advantage of the Treynor Index is that it is not dependent on the leverage and thus is a more intrinsic measure.

Property 6.3.1 The Treynor Index does not depend on leverage.

Proof Let P_λ be the portfolio of allocation $\Pi_\lambda = \lambda \Pi_P + (1 - \lambda)\Pi_0$ with $\lambda > 0$. Its Treynor Index is

$$\frac{m_{P_\lambda} - r_0}{\beta_T(P_\lambda)} = \frac{\lambda(m_P - r_0)}{\lambda \beta_T(P)} = \frac{m_P - r_0}{\beta_T(P)}. \qquad \square$$

6.4   Other Risk/Return Indicators

Other indicators are also commonly used:

- The **Sortino ratio**, which is similar to the Sharpe ratio but instead of dividing the excess return by the standard deviation of the returns, the division is made by the semi-deviation of the returns, which measures only the downside risk and not the full variability of the returns.
- The **Modigliani–Modigliani ratio** or M^2 ratio, which is defined as

$$M^2 = r_0 + \sigma_M \times \text{Sharpe ratio},$$

where σ_M is the standard deviation of the returns of the benchmark. Therefore, this ratio is the expected return that the portfolio would have if it was leveraged to have the same standard deviation as its benchmark.

- The **Information ratio**, which is used to choose between funds following the same benchmark. It is defined for a portfolio P as

$$\text{IR} = \frac{m_P - m_B}{\sigma(R_P - R_B)},$$

where m_B is the expected return of the Benchmark.

6.5   The Diversification Ratio

The Diversification ratio is an indicator which was introduced by Tasche [88], Choueifaty and Coignard [27, 28] and which summarises in a single number the benefit of diversification for a portfolio. The definition is as follows.

Definition 6.5.1 (Diversification Ratio)   If π is the risky allocation of a portfolio, different from the risk-free asset, the **Diversification ratio** for π is defined as

$$D(\pi) = \frac{\pi'\sigma}{\sigma(\pi)},$$

where $\sigma = (\sigma_1, \sigma_2, \cdots, \sigma_d)'$ is the vector of the standard deviation of the assets returns and $\sigma(\pi)$ is the standard deviation of a portfolio of risky allocation π.

In Definition 6.5.1 the portfolio is assumed to be different from the risk-free asset, in order for π to be non-zero, and for the definition to make sense. The risky investments providing the maximum diversification effect have an expression given in Proposition 6.5.1.

Proposition 6.5.1 *Let Λ be the correlation matrix for the risky assets returns, then*

$$\sup_{\pi \neq 0} D(\pi) = \sqrt{\sigma' \Sigma^{-1} \sigma} = \sqrt{1_d' \Lambda^{-1} 1_d}$$

and the sup *is reached for all risky investments of the form $\pi = \alpha \Sigma^{-1} \sigma$ with $\alpha > 0$.*

Proof $D(\pi) = \frac{\pi'\sigma}{\sigma(\pi)} = \frac{\pi'\sigma}{\sqrt{\pi'\Sigma\pi}}$. As $\pi'\sigma = \langle \Sigma\pi, \sigma \rangle_{\Sigma^{-1}}$ and $\sqrt{\pi'\Sigma\pi} = \langle \Sigma\pi, \Sigma\pi \rangle_{\Sigma^{-1}}$, geometrically the problem is to find a vector $\Sigma\pi$ which, once renormalised, maximises the scalar product with the vector σ. Therefore, the solution is any vector of the form $\alpha\sigma$ with $\alpha > 0$. From the property $\Sigma\pi = \alpha\sigma$ we get $\pi = \alpha\Sigma^{-1}\sigma$ and for such a risky allocation we get $\pi'\sigma = \alpha\sigma'\Sigma^{-1}\sigma$ and $\pi'\Sigma\pi = \alpha^2\sigma'\Sigma^{-1}\Sigma\Sigma^{-1}\sigma = \alpha^2\sigma'\Sigma^{-1}\sigma$, and consequently $D(\pi) = \sqrt{\sigma'\Sigma^{-1}\sigma}$. Now, as discussed in Proposition 3.2.2, we have $\Sigma = \mathrm{diag}(\sigma_i)\Lambda\mathrm{diag}(\sigma_i)$, so $\Sigma^{-1} = \mathrm{diag}(\frac{1}{\sigma_i})\Lambda^{-1}\mathrm{diag}(\frac{1}{\sigma_i})$ and as $\sigma'\mathrm{diag}(\frac{1}{\sigma_i}) = 1_d'$ and $\mathrm{diag}(\frac{1}{\sigma_i})\sigma = 1_d$, we get $\sigma'\Sigma^{-1}\sigma = 1_d'\Lambda^{-1}1_d$, which finishes the proof. $\square$

By comparing the value of $D(\pi)$ for a particular portfolio to the quantity $\sqrt{1_d'\Lambda^{-1}1_d}$, one can check the relative level of diversification achieved by an allocation. Note that, most likely, the Tangent Portfolio will not be amongst the portfolios offering the maximum Diversification ratio, neither the minimum variance portfolio, nor the Market Portfolio (calculated with market capitalisation weights). So, the reason to allocate into a maximum Diversification ratio portfolio are more linked to empirical results on the performance achieved than to the Markowitz paradigm. Risk Parity investing, for which the same amount of risk is allocated to each risky position, is amongst these popular new investment paradigms, aiming at reducing the risks in a robust way, but the Diversification ratio paradigm presents the advantage of taking into account the correlation, even if not in the same way as Markowitz. For an insight into these new techniques one can start with Qian [70], Choueifaty and Coignard [27] or Roncalli [75]. To get an overview of some additional new measures and ratios related to risk diversification one can find a good résumé in Fragkiskos [46].

Often, the sup for the Diversification ratio is considered only for portfolios satisfying additional constraints to the constraint $\pi \neq 0$. These constraints, for example, can be constraints of positiveness for all the components of π. In this case, no short-selling is allowed amongst the risky assets and for such portfolios the following property is satisfied.

Property 6.5.1 The following properties hold:

- $\forall \pi \in (\mathbb{R}^+)^d \setminus \{0\}$, $D(\pi) \geq 1$,
- the portfolios with a single risky asset exposure have a Diversification ratio of 1,
- the only portfolios with a minimum Diversification ratio are the portfolios with a single risky asset exposure.

Proof It is immediate that the portfolios with a single risky asset exposure have a Diversification ratio of 1. Let $\rho_{i,j}$ be the correlation between the returns of asset i and the returns of asset j. $(\pi'\boldsymbol{\sigma})^2 = \sum_{i,j=1}^{n} \pi_i \sigma_i \pi_j \sigma_j$ and $\sigma^2(\pi) = \sum_{i,j=1}^{n} \pi_i \sigma_i \pi_j \sigma_j \rho_{i,j}$ as it is assumed here that $\forall i \in [\![1,n]\!]$, $\pi_i \geq 0$, then from the fact any correlation satisfies $\rho_{i,j} \leq 1$, we get $(\pi'\boldsymbol{\sigma})^2 \geq \sigma^2(\pi)$, which proves that $D(\pi) \geq 1$. The equality between the two double summations can be reached if and only if $\pi_i \pi_j \neq 0 \implies \rho_{i,j} = 1$. As Σ is assumed to be invertible then necessarily, as demonstrated in Property 3.2.2, for all distinct indices i and j we have $\rho_{i,j} \neq 1$ and therefore $D(\pi) = 1$ implies that only one component of π is non-zero, which means that the risky investment is only in one of the risky assets. $\square$

Property 6.5.2 gives the expression of the correlation between a risky investment portfolio and a maximum Diversification ratio investment portfolio, as a function of their Diversification ratios. From there, the remarkable property that a risky investment portfolio, with maximum Diversification ratio, has the same correlation with all the individuals risky assets S_i can be shown.

Property 6.5.2 Let $\pi_\alpha = \alpha \Sigma^{-1} \boldsymbol{\sigma}$ with $\alpha > 0$ be the risky allocation of an investment portfolio with maximum Diversification ratio. Let π_P be the risky allocation of an investment portfolio P. Then,

$$\text{correl}(R(\pi_\alpha), R(\pi_P)) = \frac{D(\pi_P)}{D(\pi_\alpha)}.$$

Proof If P is the risk-free asset the correlation is zero and the formula is valid. If P is not the risk-free asset then $\pi_P \neq 0$ and we have

$$\text{correl}(R(\pi_\alpha), R(\pi_P)) = \frac{\pi'_P \Sigma \pi_\alpha}{\sigma(\pi_P)\sigma(\pi_\alpha)}.$$

As $\pi'_P \Sigma \pi_\alpha = \alpha \pi'_P \Sigma \Sigma^{-1} \boldsymbol{\sigma} = \alpha \pi'_P \boldsymbol{\sigma}$, we get

$$\text{correl}(R(\pi_\alpha), R(\pi_P)) = \alpha \frac{\pi'_P \boldsymbol{\sigma}}{\sigma(\pi_P)\sigma(\pi_\alpha)} = \alpha \frac{D(\pi_P)}{\sigma(\pi_\alpha)}. \tag{6.5.1}$$

Now, applying this formula to $\pi_P = \pi_\alpha$, we get

$$\mathrm{correl}(R(\pi_\alpha), R(\pi_\alpha)) = \alpha \frac{D(\pi_\alpha)}{\sigma(\pi_\alpha)},$$

which implies that $\alpha \frac{D(\pi_\alpha)}{\sigma(\pi_\alpha)} = 1$ and that $\alpha = \frac{\sigma(\pi_\alpha)}{D(\pi_\alpha)}$. Replacing now α in Eq. (6.5.1), we get

$$\mathrm{correl}(R(\pi_\alpha), R(\pi_P)) = \frac{\sigma(\pi_\alpha)}{D(\pi_\alpha)} \frac{D(\pi_P)}{\sigma(\pi_\alpha)} = \frac{D(\pi_P)}{D(\pi_\alpha)},$$

which finishes the proof.   □

Corollary 6.5.1 *Let* $\pi_\alpha = \alpha \Sigma^{-1} \sigma$ *with* $\alpha > 0$ *be the risky allocation of an investment portfolio with maximum Diversification ratio. Then, for any risky asset* $S_i, i \in [\![1, n]\!]$

$$\mathrm{correl}(R(\pi_\alpha), R_i) = \frac{1}{D(\pi_\alpha)}.$$

Proof For a risky asset S_i, the risky allocation is the vector π with one component equal to 1 and the others to 0. Therefore, its Diversification ratio is 1 and $\frac{D(\pi_{S_i})}{D(\pi_\alpha)} = \frac{1}{D(\pi_\alpha)}$.   □

A Few References

1. Bell, A. R., Brooks, C., & Prokopczuk, M. (2013, January 1). *Handbook of research methods and applications in empirical finance.* Cheltenham: Edward Elgar Publishing.
2. Choueifaty, Y., & Coignard, Y. (2008). Toward maximum diversification. *Journal of Portfolio Management, 35*(1), 40–51.
3. Choueifaty, Y., & Coignard, Y. (2011). Properties of the Most Diversified Portfolio, Working paper.
4. Fama, E. (1965). The behavior of stock market movements. *Journal of Business, 38*, 1749–1778.
5. Fragkiskos, A. (2014). What a CAIA member should know. *Alternative Investment Analyst Review, Q2 2014, 3*(1), 1–18.
6. Frankfurter, G. M., Phillips, H. E., & Seagle, J. P. (1974). Bias in estimating portfolio alphas and beta scores. *Review of Economics and Statistics, 56*, 412–414.
7. Guerard, J. B. (Series Editor). (2009, December 12). *Handbook of portfolio construction: Contemporary applications of Markowitz techniques.* New York: Springer.

8. Jarrow, R. A., Maksimovic, V., Ziemba, W. T. (Series Editor). *Handbook in operations research and management science* (Vol. 9). Amsterdam: Finance Elsevier.
9. Jensen, M. C. (1968). The performance of mutual funds in the period 1945–1964. *Journal of Finance, 23*, 389–416.
10. Qian, E. E. (2016). *Risk parity fundamentals* (246 pp.). New York: Chapman and Hall/CRC.
11. Modigliani, F., & Modigliani, L. (1997). Risk-adjusted performance. *The Journal of Portfolio Management, 23*(2), 45–54.
12. Roncalli, T. (2013). *Introduction to risk parity and budgeting.* Chapman and Hall/CRC Financial Mathematics Series (440 pp.). Boca Raton: CRC Press.
13. Sharpe, W. F. (1994). The Sharpe ratio. *Journal of Portfolio Management, 21*(1)(Fall), 49–58.
14. Sharpe, W. F. (1992). Asset allocation: Management style and performance measurement. *Journal of Portfolio Management, 18*(2) (Winter), 29–34.
15. Tasche, D. (2006). Measuring Sectoral Diversification in an Asymptotic Multi-Factor Framework. arXiv: physics/0505142.
16. Treynor, J. L. (1961). Toward a theory of market value of risky assets, mimeo, subsequently published in Korajczyk, Robert A. (1999). In *Asset pricing and portfolios performance: Models, strategy and performance metrics.* London: Risk Books.

Risk Measures and Capital Allocation

7

Risk measures are widely used in risk management, and to calculate capital requirements when investing or conducting banking or insurance activities. In this chapter, we study risk measures in the context of asset allocation, and explain the notions of **Value at Risk**, **Expected Shortfall** and **Return on Risk-Adjusted Capital** (**RORAC**). We provide some explicit formulas in the Gaussian framework and an example of calculation based on historical data, without any model assumptions. **Euler's formula** is presented, for standard homogeneous risk measures, as well as its applications for capital allocation between risky positions. We also prove that, when the capital is allocated according to Euler's formula, each position produces the same RORAC.

7.1   Definition of a Risk Measure

A risk measure $\mathbf{RM}(\cdot)$ is an operator on a set of random variables. The random variables L to which it applies represent the loss that an activity may generate, and the quantity $\mathbf{RM}(L)$ represents the capital necessary to conduct this activity and to be able to pare the potential losses with sufficient probability.

Definition 7.1.1 (Risk Measure) If $\mathcal{L}$ is a space of random variables with values in $\mathbb{R}$, any operator $\mathbf{RM} : \mathcal{L} \longrightarrow \mathbb{R}$ is called a **risk measure** on $\mathcal{L}$.

The convention used here is that, when L is positive a loss is made and when L is negative a gain is made. So $L = 10$ means a loss of 10 and $L = -10$ means a gain of 10.

© Springer Nature Switzerland AG 2020

P. Brugière, *Quantitative Portfolio Management*, Springer Texts in Business and Economics, https://doi.org/10.1007/978-3-030-37740-3_7

Artzner et al. in their 1999 paper [6] define four features that risk measures are usually expected to satisfy, and risk measures satisfying these properties are called **coherent**. The four features considered are the following:

Property 7.1.1 (Translation Invariance)

$$\forall L \in \mathcal{L}, \forall x \in \mathbb{R}, \mathbf{RM}(L + x) = \mathbf{RM}(L) + x.$$

This property, usually combined with the natural condition $\mathbf{RM}(0) = 0$, implies that

$$\forall x \in \mathbb{R}, \mathbf{RM}(x) = x.$$

So, in terms of capital, the interpretation is that, for example, if a loss of $x = 10$ is certain, the capital required is 10. Also, we get the intuitive property that if $\mathbf{RM}(L)$ is the capital required for a business whose loss variable is L, then by adding a deterministic loss of x to L, the capital required increases to $\mathbf{RM}(L) + x$. This translation invariance feature is satisfied for the risk measures we consider here for portfolio management.

Property 7.1.2 (Sub-additivity)

$$\forall L_1, L_2 \in \mathcal{L}, \mathbf{RM}(L_1 + L_2) \leq \mathbf{RM}(L_1) + \mathbf{RM}(L_2).$$

This property is about diversification effects. It says that less capital is required to run several businesses as a single business than to capitalise each of these businesses separately. Even if this property is satisfied for the risk measures that we use here, for portfolio management in a Gaussian law context, the property is not necessarily satisfied when applying the same risk measures to random variables modelling rare catastrophic events. In this case, the non-sub-additivity can be seen as a reason why some risk measures, used in insurance and reinsurance, encourage the creation of independent entities within the same group, as the total capital necessary to insure all these entities on a standalone basis may be less than the capital required to insure all the risks if they were aggregated.

Property 7.1.3 (Positive Homogeneity)

$$\forall L \in \mathcal{L}, \forall t > 0, \mathbf{RM}(tL) = t\mathbf{RM}(L).$$

Some people contest the desirability of this property and in some circumstances the regulation does not follow this principle. For example, in the banking industry large single equity positions, referred to as concentrated equity positions, require proportionally more capital to hold them than smaller positions in the same stocks. So, the relationship may be satisfied but only up to a certain threshold. Seen from a different angle, this homogeneity condition may seem natural, when considering for example a loss expressed in two different currencies such as USD and EUR, as it seems legitimate to expect that the value of the capital requirement does not depend on the reference currency used. This positive homogeneity feature is satisfied for the risk measures that we study in this section.

Property 7.1.4 (Monotonicity)

$$\forall L_1, L_2 \in \mathcal{L}, L_1 \geq L_2 \text{ almost surely } \Rightarrow \mathbf{RM}(L_1) \geq \mathbf{RM}(L_2).$$

This assumption seems quite logical. For two Gaussian distributions the property $L_1 \geq L_2$ implies that the two random variables differ only by a constant. The risk measures we use in this section will satisfy this property in the Gaussian law context.

7.2 Risk Measure in the Markowitz Framework

We consider a two-period model in which some risky positions are held. The risky positions are expressed as percentages of a notional x_0 and can be seen either as the risky part of a self-financing portfolio with notional x_0 or as the risky portion of an investment portfolio with initial value x_0. We define a **loss function** for this risky position, which can be seen either as the opposite of the return of the corresponding self-financing portfolio, or as the opposite of the excess return (to the risk-free asset) of the investment portfolio. Risk measures are then defined for this loss function and determine the capital required to hold these risky positions.

We assume, as always, that $M \neq r_0 1_d$ and that the matrix of variance-covariance Σ between the returns of the risky assets is invertible. We denote by Σ^{-1} its inverse, $\langle \cdot, \cdot \rangle_{\Sigma^{-1}}$ the scalar product it defines and $\| \cdot \|_{\Sigma^{-1}}$ the associated norm.

7.2.1 The Markowitz Risk Measure

Definition 7.2.1 (Loss Function) The **loss function** for the risky exposure π is defined as

$$L_\pi = -\pi'(R - r_0 1_d).$$

Remark 7.2.1

(1) $-L_\pi$ is the return of the self-financing portfolio of risky allocation π,
(2) $-L_\pi$ is the excess return (compared to the risk-free rate r_0) of the investment portfolio of risky allocation π.

Definition 7.2.2 (Markowitz Risk Measure) For the risky exposure π the **Markowitz risk measure** with parameter λ is the quantity

$$\mathbf{RM}_\lambda(L_\pi) = -\mathbf{E}(L_\pi) + \lambda\sigma(L_\pi).$$

We also write $\mathbf{RL}_\lambda(\pi) = \mathbf{RM}_\lambda(L_\pi)$.

Remark 7.2.2

$$\mathbf{RL}_\lambda(\pi) = -\pi'(M - r_0 1_d) + \lambda\sqrt{\pi'\Sigma\pi} = -\langle\Sigma\pi, M - r_0 1_d\rangle_{\Sigma^{-1}} + \lambda\|\Sigma\pi\|_{\Sigma^{-1}}.$$

Remark 7.2.3 The parameter λ can be seen as a Lagrangian parameter for a Markowitz optimisation problem and in this context can only take values above a certain threshold. This feature is detailed below.

When defining a risk measure, it is natural to impose that any self-financing portfolio requires capital (as we assume that capital cannot be created out of nothing) and therefore that $\forall\pi \in \mathbb{R} \setminus \{0\}$, $\mathbf{RL}_\lambda(\pi) > 0$. We see in the next proposition how this property limits the possible values for λ in the definition.

Definition 7.2.3 (Positive Definite Risk Measure) $\mathbf{RM}_\lambda(\cdot)$ is said to be **positive definite** if and only if $\forall\pi \in \mathbb{R} \setminus \{0\}$, $\mathbf{RL}_\lambda(\pi) > 0$.

Proposition 7.2.1 (Positive Definiteness and Parameter Range)

$$\forall\pi \in \mathbb{R} \setminus \{0\}, \mathbf{RL}_\lambda(\pi) > 0 \Leftrightarrow \lambda > \|M - r_0 1_d\|_{\Sigma^{-1}}.$$

Proof $\mathbf{RL}_\lambda(\pi) = -\langle\Sigma\pi, M - r_0 1_d\rangle_{\Sigma^{-1}} + \lambda\|\Sigma\pi\|_{\Sigma^{-1}}$. For $\pi \in \mathbb{R}^d$, we use the decomposition

$$\Sigma\pi = \beta(M - r_0 1_d) + v,$$

where v is orthogonal to $M - r_0 1_d$ with respect to $\langle\cdot, \cdot\rangle_{\Sigma^{-1}}$ and $\beta \in \mathbb{R}$.

So, $\mathbf{RL}_\lambda(\pi) = -\beta\|M - r_0 1_d\|_{\Sigma^{-1}}^2 + \lambda\sqrt{\beta^2\|M - r_0 1_d\|_{\Sigma^{-1}}^2 + \|v\|_{\Sigma^{-1}}^2}$.

Then:

- if $\lambda < \|M - r_0 1_d\|_{\Sigma^{-1}}$, by taking $v = 0$ we get $\lim\limits_{\beta \longrightarrow +\infty} \mathbf{RL}_\lambda(\pi) = -\infty$ (measure non-positive),
- if $\lambda = \|M - r_0 1_d\|_{\Sigma^{-1}}$, then if $v = 0$ and $\beta > 0$, we get $\mathbf{RL}_\lambda(\pi) = 0$ (measure non-positive definite),
- if $\lambda > \|M - r_0 1_d\|_{\Sigma^{-1}}$, then $\mathbf{RL}_\lambda(\pi) \geq 0$ and reaches its minimum value 0 only for $\pi = 0$.

So, the critical value for λ is $\lambda_0 = \|M - r_0 1_d\|_{\Sigma^{-1}}$ and only risk measures with $\lambda > \lambda_0$ will have the positive definiteness property required.   □

In the next section we present the Value at Risk and Expected Shortfall risk measures, which are particular cases of the Markowitz risk measure, when working under the Gaussian framework.

7.2.2 Value at Risk

We denote by Φ the cumulative distribution function for a normal variable $Z \sim \mathcal{N}(0, 1)$ and ϕ its derivative, i.e. the density. The Value at Risk was introduced by the bank JP Morgan in 1993 and is defined in the following way.

Definition 7.2.4 (Value at Risk) For $\alpha \in {]}0, 1[$, the risk measure

$$\mathbf{VaR}_\alpha(L_\pi) = \inf\{x \in \mathbb{R}, P(L_\pi \leq x) \geq \alpha\}$$

is called the **Value at Risk** at level α.

We have the following property.

Property 7.2.1 In the Markowitz framework,

$$P(L_\pi \leq \mathbf{VaR}_\alpha(L_\pi)) = \alpha.$$

Proof L_π is a Gaussian variable and for a normal distribution the repartition function is a bijection from $\mathbb{R}$ to $]0, 1[$. So, $\exists!\beta$ such that $P(L_\pi \leq \beta) = \alpha$ and it is easy to check that as a consequence $\mathbf{VaR}_\alpha(L_\pi) = \beta$. □

Corollary 7.2.1 *In the Markowitz framework,*

$$\mathbf{VaR}_\alpha(L_\pi) = \mathbf{E}(L_\pi) + \Phi^{-1}(\alpha)\sigma(L_\pi)$$

and therefore the Value at Risk is a particular case of a Markowitz risk measure.

Proof L_π is a Gaussian variable and we can write $L_\pi \sim \mathbf{E}(L_\pi) + \sigma(L_\pi)Z$ where $Z \sim \mathcal{N}(0, 1)$. Therefore

$$P(L_\pi \leq \mathbf{VaR}_\alpha(L_\pi)) = \alpha$$

$$\Leftrightarrow P\big(\mathbf{E}(L_\pi) + \sigma(L_\pi)Z \leq \mathbf{VaR}_\alpha(L_\pi)\big) = \alpha$$

$$\Leftrightarrow P\Big(Z \leq \frac{1}{\sigma(L_\pi)}(-\mathbf{E}(L_\pi) + \mathbf{VaR}_\alpha(L_\pi))\Big) = \alpha$$

$$\Leftrightarrow \Phi\Big(\frac{1}{\sigma(L_\pi)}(-\mathbf{E}(L_\pi) + \mathbf{VaR}_\alpha(L_\pi))\Big) = \alpha$$

$$\Leftrightarrow \mathbf{VaR}_\alpha(L_\pi) = \mathbf{E}(L_\pi) + \Phi^{-1}(\alpha)\sigma(L_\pi).$$

□

The Value at Risk is widely used in banking and insurance. Usually, when analysing a trading book under the Basel III framework, the Value at Risk considered is for a horizon of 1 day and a parameter α of 99% (see BIS 2016 [13]). This way, if a trading desk has a **VaR** limit of EUR 10 m, risk managers expect that, when observing over 1 year the daily losses or gains of the trading desk, the number of daily losses observed above the EUR 10 m threshold will represent no more than 1% of the observations. Of course, the frequency derived from the observations may be slightly different from its theoretical value, as an estimate from a sample can differ from its theoretical value (especially when estimating rare events), but, any large deviation from the expected value should be analysed. In insurance, under Solvency II, the Values at Risk usually considered are for a horizon of 1 year and for a parameter α of 99.5% (see Boonen [22]).

7.2.3 Expected Shortfall

Another widely used risk measure is the Expected Shortfall. One of the reasons to introduce this measure is that, in contrast to the Value at Risk, which may be non-sub-additive in a non-Gaussian framework, the Expected Shortfall is always a coherent measure and therefore sub-additive (see Boonen [22]).

Definition 7.2.5 (Expected Shortfall) For $\alpha \in\,]0, 1[$ we define

$$\mathbf{E}_\alpha(L_\pi) = \mathbf{E}\big(L_\pi | L_\pi \geq \mathbf{VaR}_\alpha(L_\pi)\big)$$

and call this risk measure the **Expected Shortfall** at level α for the risk exposure π.

Lemma 7.2.1 $\forall a \in \mathbb{R}$ and $Z \sim \mathcal{N}(0, 1)$ we have

$$\mathbf{E}(Z | Z \geq a) = \frac{\phi(a)}{1 - \Phi(a)}.$$

Proof

$$\mathbf{E}(Z | Z \geq a) = \int\limits_{a}^{+\infty} z \frac{1}{\sqrt{2\pi}} \exp(-\frac{z^2}{2}) \frac{1}{1 - \Phi(a)} dz$$

$$= \frac{1}{1 - \Phi(a)} \left[-\frac{1}{\sqrt{2\pi}} \exp(-\frac{z^2}{2}) \right]_{a}^{+\infty} = \frac{\phi(a)}{1 - \Phi(a)}. \qquad \square$$

Proposition 7.2.2 *In the Markowitz framework,*

$$\mathbf{E}_\alpha(L_\pi) = \mathbf{E}(L_\pi) + \frac{\phi(\Phi^{-1}(\alpha))}{1 - \alpha} \sigma(L_\pi).$$

Proof From Lemma 7.2.1, if $X \sim \mathcal{N}(m, \sigma^2)$ then

$$\mathbf{E}(X|X \geq a) = \mathbf{E}(m + \sigma Z|m + \sigma Z \geq a)$$

$$= m + \sigma \mathbf{E}\left(Z|Z \geq \frac{a-m}{\sigma}\right)$$

$$= m + \frac{\phi(\frac{a-m}{\sigma})}{1 - \Phi(\frac{a-m}{\sigma})}\sigma.$$

Here, $L_\pi \sim \mathcal{N}\left(\mathbf{E}(L_\pi), \sigma^2(L_\pi)\right)$ so, if we define $m = \mathbf{E}(L_\pi)$ and $\sigma = \sigma(L_\pi)$, we get

$$\mathbf{E}_\alpha(L_\pi) = m + \frac{\phi(\frac{a-m}{\sigma})}{1 - \Phi(\frac{a-m}{\sigma})}\sigma,$$

with $a = m + \Phi^{-1}(\alpha)\sigma$ according to the result on the **VaR**. So

$$\mathbf{E}_\alpha(L_\pi) = m + \frac{\phi(\Phi^{-1}(\alpha))}{1 - \Phi(\Phi^{-1}(\alpha))}\sigma = m + \frac{\phi(\Phi^{-1}(\alpha))}{1 - \alpha}\sigma. \qquad \square$$

So, in the Markowitz framework the Expected Shortfall is a particular case of a Markowitz risk measure.

Remark 7.2.4 From the previous results, it follows that in a Gaussian framework an Expected Shortfall of parameter β is the same as a Value at Risk of parameter α as long as

$$\Phi^{-1}(\alpha) = \frac{\phi(\Phi^{-1}(\beta))}{1 - \beta}, \text{ i.e. } \alpha = \Phi\left(\frac{\phi(\Phi^{-1}(\beta))}{1 - \beta}\right).$$

For example, for $\beta = 97.5\%$ we get $\alpha = 99.03\%$.

Note that in Basel III, and Solvency II, the calculation of an Expected Shortfall at level 97.5% is presented as an alternative to the **VaR**$_{99\%}$. Indeed, as seen above, the two lead to similar results when applied to a normal distribution.

Note that, even if the Expected Shortfall is a coherent measure, it has its own inconveniences, which is the reason why it is not a systematic winner over the **VaR**. Other alternative risk measures are currently being explored, such as the expectiles (see Cai [24]), but are not within the scope of this book.

7.3   Euler's Formula and Capital Allocation

In this section, we calculate Markowitz risk measures and demonstrate that, in a Gaussian law context, they reward diversification, as the capital required for a risky portfolio appears to be less than the sum of the capital requirements for each of its risky components. Then, we see how Euler's formula allocates the capital efficiently in portfolio management.

For $\pi \in \mathbb{R}^d$ we write $\pi = \sum_{i=1}^{d} \pi^i e_i$ in the canonical basis $(e_i)_{i \in [\![1,d]\!]}$ of $\mathbb{R}^d$.

Proposition 7.3.1 (Decomposition of the Risk Measure) *The Markowitz risk measure* $\mathbf{RL}_\lambda(\pi)$ *has the following properties:*

(1) $\forall \pi \in \mathbb{R}^d$, $\mathbf{RL}_\lambda(\pi) \leq \sum_{i=1}^{d} \mathbf{RL}_\lambda(\pi^i e_i)$ *(diversification effect),*

(2) $\forall \pi \in \mathbb{R}^d$, $\mathbf{RL}_\lambda(\pi) = \sum_{i=1}^{d} \pi^i \frac{\partial \mathbf{RL}_\lambda}{\partial e_i}(\pi)$ *(Euler's formula).*

Proof

For (1) we have $\mathbf{RL}_\lambda(\pi) = -\langle \Sigma\pi, M - r_0 1_d \rangle_{\Sigma^{-1}} + \lambda \|\Sigma\pi\|_{\Sigma^{-1}}$, so

$$\mathbf{RL}_\lambda(\pi^i e_i) = -\langle \Sigma\pi^i e_i, M - r_0 1_d \rangle_{\Sigma^{-1}} + \lambda \|\Sigma\pi^i e_i\|_{\Sigma^{-1}} \text{ and}$$

$$\sum_{i=1}^{d} \mathbf{RL}_\lambda(\pi^i e_i) = -\sum_{i=1}^{d} \langle \Sigma\pi^i e_i, M - r_0 1_d \rangle_{\Sigma^{-1}} + \lambda \sum_{i=1}^{d} \|\Sigma\pi^i e_i\|_{\Sigma^{-1}}$$

$$= -\langle \Sigma\pi, M - r_0 1_d \rangle_{\Sigma^{-1}} + \lambda \sum_{i=1}^{d} \|\Sigma\pi^i e_i\|_{\Sigma^{-1}}.$$

Therefore the result follows according to the triangle inequality applied to the norm $\| \cdot \|_{\Sigma^{-1}}$.

RL_λ For (2) we observe that $RL_\lambda(\pi)$ is positive homogeneous of degree 1 as $\forall t \geq 0$, $R_\lambda(t\pi) = t RL_\lambda(\pi)$. So, from Euler's formula, applied to homogeneous functions of degree 1, the result follows.   □

Remark 7.3.1 The proposition above can be interpreted in the following way: the capital required to take the exposure π^i to the asset S_i on a standalone basis is $\mathbf{RL}_\lambda(\pi^i e_i)$ but when investing in several assets $\pi^i \frac{\partial \mathbf{RL}_\lambda}{\partial e_i}(\pi)$ can be interpreted as the cost of capital for taking the position π^i in asset S_i, adjusted by the diversification effect.

Corollary 7.3.1 (Single Position)

$$\forall \pi \in \mathbb{R}^d, \forall i \in [\![1, d]\!], \mathbf{RL}_\lambda(\pi^i e_i) = \pi^i \frac{\partial \mathbf{RL}_\lambda}{\partial e_i}(\pi^i e_i).$$

Proof Applying the formula $\mathbf{RL}_\lambda(\pi) = \sum_{j=1}^{d} \pi^j \frac{\partial \mathbf{RL}_\lambda}{\partial e_j}(\pi)$ to $\pi = \pi^i e_i$ gives the result, as in the sum only the term $\pi^i \frac{\partial \mathbf{RL}_\lambda}{\partial e_i}(\pi^i e_i)$ is not zero. □

Corollary 7.3.1 emphasises that Euler's formula gives, of course, in the absence of diversification, the same capital requirement as the risk measure $\mathbf{RL}_\lambda(\cdot)$ it derives from.

Remark 7.3.2 We may consider functions $\mathbf{RL}_\lambda(\pi^i e_i, \pi)$ other than $\pi_i \frac{\partial RL_\lambda}{\partial e_i}(\pi)$ which satisfy

$$\sum_{i=1}^{d} \mathbf{RL}_\lambda(\pi^i e_i, \pi) = \mathbf{RL}_\lambda(\pi) \text{ and } \mathbf{RL}_\lambda(\pi^i e_i, \pi^i e_i) = \mathbf{RL}_\lambda(\pi^i e_i)$$

and call such a function a **capital allocation** for the risk measure $\mathbf{RL}_\lambda(\cdot)$, but Euler's formula features some properties, that we describe in the next section, which makes it particularly interesting.

7.3.1 Example of Risk Measure and Capital Allocation

Let S_1 and S_2 be two risky assets of annual expected returns $m_1 = 5\%$ and $m_2 = 7\%$, of correlation zero and with standard deviations for the returns $\sigma_1 = 10\%$ and $\sigma_2 = 20\%$. We assume a risk-free rate $r_0 = 1\%$.

We assume a EUR 50 m exposure to asset 1 and EUR 50 m exposure to asset 2. We use the Markowitz risk measure $\mathbf{RL}_\lambda(\pi) = -\pi'(M - r_0 1_2) + \lambda\sqrt{\pi'\Sigma\pi}$. We take for the notional $x_0 = $ EUR 100 m, so $\pi_1 = 0.5$ and $\pi_2 = 0.5$. Then we have the following results:

(a) Risk parameters:

   (1) the matrix of variance-covariance for the returns is $\Sigma = \begin{pmatrix} 0.01 & 0 \\ 0 & 0.04 \end{pmatrix}$,

   (2) for the risk measure to always be positive we need $\lambda > \lambda_0$ with $\lambda_0 = \|M - r_0 1_2\|_{\Sigma^{-1}}$, which gives $\lambda_0 = 0.5$,

   (3) a 95% **VaR** for a 1 year horizon corresponds to $\lambda = \Phi^{-1}(.95) = 1.64$,

   (4) a 95% Expected Shortfall for a 1 year horizon corresponds to $\lambda = \frac{\phi(\Phi^{-1}(.95))}{1 - 0.95} = 2.06$.

(b) Capital required for the risky positions aggregated under $\mathbf{VaR}_{95\%}$:
  (1) $\pi'(M - r_0 1_2) = 0.5 \times (5\% - 1\%) + 0.5 \times (7\% - 1\%) = 5\%$,
  (2) $\sqrt{\pi'\Sigma\pi} = 11.18\%$,
  (3) $\mathbf{VaR}_{95\%} = -5\% + 1.64 \times 11.18\% = 13.39\%$, so EUR 13.39 m.
(c) Capital required for each position on a standalone basis under $\mathbf{VaR}_{95\%}$:
  (1) for asset 1: $-0.5 \times (5\% - 1\%) + 1.64 \times 0.5 \times 10\% = 6.20\%$, so EUR 6.20 m,
  (2) for asset 2: $-0.5 \times (7\% - 1\%) + 1.64 \times 0.5 \times 20\% = 13.40\%$, so EUR 13.40 m.

  So, the diversification effect, and sub-additivity property in a Gaussian law context, is visible here for the $\mathbf{VaR}$, as the sum of the capital requirements, calculated on a standalone basis, is EUR 19.60 m, which is much higher than the EUR 13.39 m calculated on an aggregated basis.
(d) We now use Euler's formula to allocate the EUR 13.34 m capital requirement between the two positions.

$$\mathbf{RL}_\lambda(\pi) = -\pi'(M - r_0 1_2) + \lambda\sqrt{\pi'\Sigma\pi} \implies \frac{\partial \mathbf{RL}_\lambda}{\partial e_i}(\pi) = -(m_i - r_0) + \frac{\lambda}{\sqrt{\pi'\Sigma\pi}}\langle\Sigma\pi, e_i\rangle.$$ So,
  (1) for the first asset $\pi_1 \frac{\partial \mathbf{RL}_\lambda}{\partial e_1}(\pi) = -0.5 \times (5\% - 1\%) + 0.5 \times \frac{1.64}{11.18\%} \times 0.01 = 1.68\%$, so EUR 1.68 m,
  (2) for the second asset $\pi_2 \frac{\partial \mathbf{RL}_\lambda}{\partial e_2}(\pi) = -0.5 \times (7\% - 1\%) + 0.5 \times \frac{1.64}{11.18\%} \times 0.04 = 11.71\%$, so EUR 11.71 m,
  (3) when we add the Euler capital allocations we get EUR 1.68 m + EUR 11.71 m = EUR 13.39 m, which is the expected result.

  In the next section we present an additional property of the Euler capital allocation when calculating the Return on Risk-Adjusted Capital.

7.4 Return on Risk-Adjusted Capital

7.4.1 Maximising the RORAC

First we start by proving that for a positive definite Markowitz risk measure, maximising the RORAC for an allocation between risky assets is the same as maximising the Sharpe ratio for the corresponding investment portfolios.

Definition 7.4.1 (Return on Risk-Adjusted Capital) For a risky asset allocation π and risk measure $\mathbf{RL}_\lambda(\pi)$, we define the **RORAC** by

$$\mathrm{RORAC}_\lambda(\pi) = \frac{-\mathrm{E}(L_\pi)}{\mathbf{RL}_\lambda(\pi)}.$$

We demonstrate in Proposition 7.4.1 that if $\mathbf{RL}_\lambda(\cdot)$ is a positive definite Markowitz risk measure (i.e. $\lambda > \lambda_0$), then the risky allocations π which maximise the RORAC correspond to the risky allocations of the investment portfolios on the Capital Market Line.

Proposition 7.4.1 *In a Markowitz model, if we consider a Markowitz risk measure* $\mathbf{RL}_\lambda(\pi)$ *with* $\lambda > \|M - r_0 1_d\|_{\Sigma^{-1}}$ *then the solutions* π *to the optimisation problem*

$$\sup_{\pi \in \mathbb{R}^d \setminus \{0\}} \frac{-\mathbf{E}(L_\pi)}{\mathbf{RL}_\lambda(\pi)}$$

correspond to the risky allocations of the investment portfolios on the Capital Market Line.

Proof Let $R(\pi)$ be the return of the investment portfolio of risky allocation π. As seen previously, the condition $\lambda > \|M - r_0 1_d\|_{\Sigma^{-1}}$ implies that $\mathbf{RL}_\lambda(\pi) > 0$, so the ratio $\mathbf{RL}_\lambda(\pi)$ is well defined for all $\pi \in \mathbb{R}^d \setminus \{0\}$.

Now,

$$\arg\max_{\pi \in \mathbb{R}^d \setminus \{0\}} \mathrm{RORAC}_\lambda(\pi) = \arg\max_{\pi \in \mathbb{R}^d \setminus \{0\}} \frac{\mathbf{E}(R(\pi) - r_0)}{-\mathbf{E}(R(\pi) - r_0) + \lambda\sigma(R(\pi))}$$

$$= \arg\max_{\pi \in \mathbb{R}^d \setminus \{0\}} \frac{\frac{\mathbf{E}(R(\pi) - r_0)}{\lambda\sigma(R(\pi))}}{-\frac{\mathbf{E}(R(\pi) - r_0)}{\lambda\sigma(R(\pi))} + 1}.$$

So we search for

$$\arg\max_{x(\pi) = \frac{\mathbf{E}(R(\pi) - r_0)}{\lambda\sigma(R(\pi))} < 1} \frac{x(\pi)}{1 - x(\pi)}.$$

As $\frac{\mathbf{E}(R(\pi) - r_0)}{\lambda\sigma(R_\pi)} < 1$ and $\exists\pi$ such that $\frac{\mathbf{E}(R(\pi) - r_0)}{\lambda\sigma(R(\pi))} > 0$, the maximum will be positive and will be obtained for values of $x(\pi)$ as close as possible to 1.
So,

$$\arg\max_{\pi \in \mathbb{R}^d \setminus \{0\}} \mathrm{RORAC}_\lambda(\pi) = \arg\max_{\pi \in \mathbb{R}^d \setminus \{0\}} \frac{\mathbf{E}(R(\pi) - r_0)}{\lambda\sigma(R(\pi))}.$$

Therefore, the risky allocations which maximise the RORAC are the risky allocations, of the investment portfolios, which maximise the Sharpe Ratio, and therefore correspond to the risky allocations of the investment portfolios on the Capital Market Line. □

7.4.2   Capital Allocation for a Positive Homogeneous Risk Measure

In this section, we study how the total capital required for a risky exposure π can be allocated between its risky components $\pi^i e_i$. So, we study risk measures $\mathbf{RL}(\cdot)$ and risk allocation functions $\mathbf{RL}(\cdot, \cdot)$ such that for any risky portfolio π,

$$\mathbf{RL}(\pi) = \sum_{i=1}^{d} \mathbf{RL}(\pi^i e_i, \pi) \text{ and } \mathbf{RL}(\pi^i e_i) = \mathbf{RL}(\pi^i e_i, \pi^i e_i).$$

So, the capital allocation function $\mathbf{RL}(\cdot, \cdot)$ gives the same capital requirement as the risk measure $\mathbf{RL}(\cdot)$ for single risky positions (with therefore no diversification effect) and, in the case of multiple risky positions, $\mathbf{RL}(\cdot, \pi)$ allocates the capital required $\mathbf{RL}(\pi)$ between all the risky positions, taking into account the diversification effect for each of the risky positions.

Definition 7.4.2 (marginal RORAC) If $\mathbf{RL}(\cdot, \cdot)$ is a capital allocation function for the risk measure $\mathbf{RL}(\cdot)$, the **marginal RORAC** of the position $\pi^i e_i \neq 0$ in the portfolio π for the capital allocation function $\mathbf{RL}(\cdot, \cdot)$ is defined as

$$\mathrm{RORAC}(\pi^i e_i, \pi) = \frac{\pi^i (m_i - r_0)}{\mathbf{RL}(\pi^i e_i, \pi)}.$$

Definition 7.4.3 (RORAC-Compatible   Allocations) $\mathbf{RL}(\cdot, \cdot)$ is   **RORAC-compatible** with the positive definite risk measure $\mathbf{RL}(\cdot)$ it relates to iff property $(P1)$ implies property $(P2)$, where

$$(P1) : \mathrm{RORAC}(\pi^i e_i, \pi) > \mathrm{RORAC}(\pi) \text{ and } \pi^i \mathbf{RL}(\pi^i e_i, \pi) > 0, \qquad (7.4.1)$$

$$(P2) : \exists \epsilon > 0, \forall h \in]0, \epsilon[\mathrm{RORAC}(\pi + h\pi^i e_i) > \mathrm{RORAC}(\pi). \qquad (7.4.2)$$

Proposition 7.4.2 *If* $\mathbf{RL}(\cdot)$ *is a positive definite risk measure which is differentiable on* $\mathbb{R}^d \setminus \{0\}$ *and positive homogeneous and if* $\mathbf{RL}(\cdot, \cdot)$ *is its Euler capital allocation function then* $\mathbf{RL}(\cdot, \cdot)$ *is RORAC-compatible with* $RL(.)$.

Proof As $\mathbf{RL}(\cdot)$ is assumed to be positive homogeneous, Euler's formula defines a capital allocation function

$$\pi^i \mathbf{RL}(\pi^i e_i, \pi) = (\pi^i)^2 \frac{\partial \mathbf{RL}}{\partial e_i}(\pi).$$

So, $\pi^i \mathbf{RL}(\pi^i e_i, \pi) > 0 \Rightarrow \frac{\partial \mathbf{RL}}{\partial e_i}(\pi) > 0$.
    Also,

$$\mathrm{RORAC}(\pi^i e_i, \pi) > \mathrm{RORAC}(\pi) \Rightarrow \frac{\pi^i (m_i - r_0)}{\pi^i \frac{\partial \mathbf{RL}}{\partial e_i}(\pi)} > \frac{\pi'(M - r_0 1_d)}{\mathbf{RL}(\pi)},$$

which implies that

$$\frac{(m_i - r_0)}{\frac{\partial \mathbf{RL}}{\partial e_i}(\pi)} > \frac{\pi'(M - r_0 1_d)}{\mathbf{RL}(\pi)}. \qquad (7.4.3)$$

To show the property it is enough to show that $\frac{\partial \mathrm{RORAC}}{\partial e_i}(\pi) > 0$.

As

$$\frac{\partial \, \text{RORAC}}{\partial e_i}(\pi) = \frac{(m_i - r_0)\mathbf{RL}(\pi) - \pi'(M - r_0 1_d)\frac{\partial \mathbf{RL}}{\partial e_i}(\pi)}{(\mathbf{RL}(\pi))^2}, \qquad (7.4.4)$$

this results from Eq. (7.4.3) since by assumption $\mathbf{RL}(\pi)$ and $\frac{\partial \mathbf{RL}}{\partial e_i}(\pi)$ are both strictly positive.   □

Corollary 7.4.1 *If* $\mathbf{RL}_\lambda(\cdot)$ *is a positive definite Markowitz risk measure (i.e.* $\lambda >$ λ_0*) then*

$$\mathbf{RL}_\lambda(\pi^i e_i, \pi) = \pi^i \frac{\partial \mathbf{RL}_\lambda}{\partial e_i}(\pi)$$

is a capital allocation function for $\mathbf{RL}_\lambda(\cdot)$ *which is RORAC-compatible.*

Proof $\mathbf{RL}_\lambda(\cdot)$ satisfies the conditions of Proposition 7.4.2, which proves the result.   □

Proposition 7.4.3 *Let* $\mathbf{RL}(\cdot)$ *be a positive definite Markowitz risk measure, homogeneous and differentiable on* $\mathbb{R}^d \setminus \{0\}$*. If a risky allocation* $\pi^* \neq 0$ *maximises the RORAC then for all risky positions the individual RORAC of the position* $\pi^{*i} e_i$*, calculated with the Euler capital allocation* $\pi^{*i} \frac{\partial \mathbf{RL}}{\partial e_i}(\pi^*)$*, equals the RORAC of the portfolio if* $m_i \neq r_0$*.*

Proof The RORAC of the portfolio is $\frac{\pi'(M - r_0 1_d)}{\mathbf{RL}(\pi)}$. When the RORAC is at its maximum the derivative given in Eq. (7.4.4) cancels, which implies that

$$(m_i - r_0)\mathbf{RL}(\pi^*) = \pi^{*\prime}(M - r_0 1_d)\frac{\partial \mathbf{RL}}{\partial e_i}(\pi^*).$$

The left term is different from zero as by assumption $m_i \neq r_0$ and $\mathbf{RL}$ is positive definite. So the terms are not zero and we get $\frac{(m_i - r_0)}{\frac{\partial \mathbf{RL}}{\partial e_i}(\pi^*)} = \frac{\pi^{*\prime}(M - r_0 1_d)}{\mathbf{RL}(\pi^*)}$ and the result.   □

Proposition 7.4.4 (Maximum RORAC) *For a positive definite Markowitz risk measure*

$$R_\lambda(\pi) = -\pi'(M - r_0 1_d) + \lambda\sqrt{\pi'\Sigma\pi}$$

the maximum RORAC is $\frac{1}{\frac{\lambda}{\lambda_0} - 1}$*, where* λ_0 *is the critical value* $\|M - r_0 1_d\|_{\Sigma^{-1}}$*.*

Proof $\mathrm{RORAC}(\pi) = \frac{\pi'(M-r_0 1_d)}{\mathbf{RL}_\lambda(\pi)}$ and we know from Proposition 7.4.1 that the maximum is reached on the Capital Market Line, so we can calculate this quantity for $\pi^* = \Sigma^{-1}(M - r_0 1_d)$ and get

$$\pi^{*\prime}(M - r_0 1_d) = \|M - r_0 1_d\|^2_{\Sigma^{-1}}, \text{ and}$$

$$\pi^{*\prime}\Sigma\pi^* = \|M - r_0 1_d\|^2_{\Sigma^{-1}}, \text{ so}$$

$$\mathbf{RL}_\lambda(\pi^*) = -\|M - r_0 1_d\|^2_{\Sigma^{-1}} + \lambda\|M - r_0 1_d\|_{\Sigma^{-1}} \text{ and}$$

$$\mathrm{RORAC}(\pi^*) = \frac{\lambda_0^2}{-\lambda_0^2 + \lambda\lambda_0},$$

which proves the result.   □

7.4.3 Example: Euler Allocation

We consider the same risky assets S_1 and S_2 as in Example 7.3.1, with annual expected returns $m_1 = 5\%$ and $m_2 = 7\%$, correlation zero and standard deviations for the returns $\sigma_1 = 10\%$ and $\sigma_2 = 20\%$, and we assume a risk-free rate $r_0 = 1\%$. We consider various allocations with π_1 varying from -100 to 200% and $\pi_2 = 1 - \pi_1$. The notional is still EUR 100 m. For all the risky portfolios considered we calculate the **VaR**$_{99\%}$ as a percentage of the Notional and the Euler allocations for asset 1 and asset 2. We verify in Fig. 7.1 that the Euler allocations add up to the

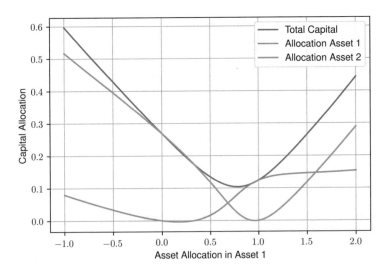

Fig. 7.1 **VaR**$_{99\%}$ allocation for the risk exposure $(\pi_1, 1 - \pi_1)$

total capital required for the portfolios. We also notice that the Euler allocations can sometimes take slightly negative values, which is something normal as the Euler allocation is not supposed to satisfy the same property of positive definiteness as the Markowitz risk measure it derives from.

7.4.4 Example: RORAC for Optimal Portfolios

We consider the same risky assets as in Example 7.4.3, but this time search risky allocations which maximise the RORAC and calculate the RORAC of these risky portfolios as well as the RORAC of their individual components, with the diversification effects taken into account. The risk measure considered is a positive definite Markowitz risk measure $\mathbf{RL}_\lambda(\cdot)$ and the capital allocation component by component is given by Euler's formula. For the measure to be positive definite λ is assumed to be strictly higher than λ_0, which is here equal to 0.5. Here we take $\lambda = 1.64$, which corresponds to a $\mathbf{VaR}_{99\%}$ in a Gaussian law context. The notional is still $x_0 = \text{EUR } 100\,\text{m}$.

As the portfolios on the Capital Market Line maximise the RORAC we choose such a portfolio with $\pi^* = \Sigma^{-1}(M - r_0 1_2)$ and obtain $\pi_1^* = 4$ and $\pi_2^* = 1.5$. Then we obtain the following results:

(1) For the portfolio π^*, we get for the RORAC

$$\pi^{*\prime}(M - r_0 1_2) = 25\%, \quad \sqrt{\pi^{*\prime}\Sigma\pi^*} = 50\%, \quad \mathbf{RL}_\lambda(\pi^*) = -0.25 + 0.5\lambda$$

and consequently,

$$\text{RORAC}_\lambda(\pi^*) = \frac{0.25}{-0.25 + 0.5\lambda} = \frac{1}{2\lambda - 1},$$

which is equal to $\frac{1}{\frac{\lambda}{\lambda_0} - 1}$, as expected.

(2) For the single position π_1^* in asset 1 we have for the marginal RORAC

$$\mathbf{RL}_\lambda(\pi_1^* e_1, \pi^*) = \pi_1^* \frac{\partial \mathbf{RL}_\lambda}{\partial e_1}(\pi^*) = -\pi_1^*(m_1 - r_0) + \pi_1^* \frac{\lambda}{\sqrt{\pi^{*\prime}\Sigma\pi^*}} \langle \Sigma\pi^*, e_1 \rangle,$$

with $\pi_1^*(m_1 - r_0) = 16\%$ and $\langle \Sigma\pi^*, e_1 \rangle = 4\%$.

So the capital allocated according to Euler's formula is

$$\mathbf{RL}_\lambda(\pi_1^* e_1, \pi^*) = -16\% + 4 \times \frac{\lambda}{0.5} \times 4\% = -16\% + \lambda \times 32\%,$$

which gives

$$\text{RORAC}_\lambda(\pi_1^* e_1, \pi^*) = \frac{0.16}{-0.16 + 0.32\lambda} = \frac{1}{2\lambda - 1},$$

which is the same as the RORAC of the portfolio π^*, as expected.

(3) For the single position π_2^* in asset 2 we have for the marginal RORAC

$$\mathbf{RL}_\lambda(\pi_2^* e_2, \pi^*) = \pi_2^* \frac{\partial \mathbf{RL}_\lambda}{\partial e_2}(\pi^*) = -\pi_2^*(m_2 - r_0) + \pi_2^* \frac{\lambda}{\sqrt{\pi^{*\prime} \Sigma \pi^*}} \langle \Sigma \pi^*, e_2 \rangle,$$

with $\pi_2^*(m_2 - r_0) = 9\%$ and $\langle \Sigma \pi^*, e_2 \rangle = 6\%$.

So the capital allocated according to Euler's formula is

$$\mathbf{RL}_\lambda(\pi_2^* e_2 \pi^*) = -9\% + 1.5 \times \frac{\lambda}{0.5} \times 6\% = -9\% + \lambda \times 18\%,$$

which gives

$$\text{RORAC}_\lambda(\pi_2^* e_2, \pi^*) = \frac{0.09}{-0.09 + 0.18\lambda} = \frac{1}{2\lambda - 1},$$

which is the same as the RORAC of the portfolio π^*, as expected.

The results obtained, for $\lambda = 1.64$, are summarised in Table 7.1 in EUR and as a percentage of $x_0 = \text{EUR } 100\,\text{m}$, for two portfolios and their components. x_0 can be interpreted either as the notional of the self-financing portfolio with these risky positions or as the initial value of the investment portfolio with these risky positions. For the optimal portfolio the RORAC of the portfolio and of its components, based on Euler's formula, are the same.

Table 7.1 RORAC for some portfolios and their components

Risk position π	$\pi'(M - r_0 1_2)$	$\sigma(R(\pi))$	Sharpe ratio	Euler capital	RORAC
$\pi = (0.5, 0.5)$	5%	11.18%	0.45	13.39 m	37.34%
$\pi_1 e_1$	2%	5%	0.40	1.68 m	119.05%
$\pi_2 e_2$	3%	10%	0.30	11.71 m	25.62%
$\pi^* = (4, 1.5)$	25%	50%	0.50	57.24 m	43.67%
$\pi_1^* e_1$	16%	40%	0.40	36.64 m	43.67%
$\pi_2^* e_2$	9%	30%	0.30	20.61 m	43.67%

7.4.5   Calculation of a Portfolio VaR, from Observed Asset Prices

In practice, the law of the asset returns is unknown and must be inferred from the observations of the asset prices, from which the vectors of the daily returns can be calculated. The **VaR** of the portfolio can then be calculated using the estimated distribution function inferred from these daily returns. The distribution function estimate can be worked out for both **parametric** and **non-parametric models**. For parametric models, the law of the asset returns is assumed to be defined by a fixed number of parameters. In the case of the normal distribution, for example, the parameters (totally) defining the model are the vector M of the asset expected returns and the variance-covariance matrix Σ. In this case, from the observations of the asset prices, M and Σ are estimated and from there, by linear transformation, the law for the portfolio returns. In this case, the estimated law for the returns of the portfolio π is a normal law of expectation $\pi' \hat{M}$ and variance $\pi' \hat{\Sigma} \pi$, where $\hat{M}$ is the estimate of M and $\hat{\Sigma}$ is the estimate of Σ, based on the observations. Using this estimated law, for the portfolio returns, the value of the **VaR** of the portfolio can be calculated explicitly, according to the formula derived in Corollary 7.2.1 for a normal distribution.

In the case of a parametric distribution, for the asset returns, which is not normal, the law for the portfolio returns may be a complex function of the parameters, and it may be impossible to find an analytic expression of the **VaR** of the portfolio based on these parameters (and their estimates). In this case, a **Monte Carlo method** can be considered, for the asset returns and consequently for the portfolio returns, by using the estimated parameters for the distribution, and from there running simulations. Once done, the (empirical) quantile corresponding to the **VaR** of the portfolio can be calculated.

In a non-parametric model, usually no assumption is made on the law of the asset returns. In this case, the calculation of the portfolio **VaR** is made by running simulations based on the empirical law of the asset returns. This empirical law is derived from the observations in the following way: if $R(i) = (r_1(i), r_2(i), \cdots, r_d(i))$ is the vector of the observed returns for the d assets over the ith period, the empirical distribution of the random variable R, of the asset returns, sets a probability of occurrence of $\frac{1}{n}$ to each realisation $R(i)$. Performing a Monte Carlo simulation based on this underlying empirical distribution of R consists in picking successively (with replacements) some realisations $R(i)$, each time with probability $\frac{1}{n}$. This process, of running a Monte Carlo simulation on the empirical distribution, is called the **bootstrap method**. Once s observations $R(i_1), R(i_2), \cdots R(i_s)$ are picked, the corresponding returns of the portfolio are calculated and, from this set of s returns, the wanted quantile can be calculated and thus the **VaR**. Calculating the **VaR** this way is often called **Bootstrap Historical Simulation** of the **VaR** or **BHS VaR**.

There is an important literature on the different ways to calculate the **VaR** and the pros and cons of each of the methods. For a detailed overview, one can consult, for example, Dowd [37], Engle and Manganelli [41] or Abada et al. [1]. In the parametric case, the problem of estimating the parameters can be complex, when working outside of the Gaussian framework, as a very wide range of models can be considered such as: **GARCH** models (which encapsulate some correlations between the variance of the returns over time), **Copula** models (which take into account some non-linear correlations between the returns of the assets) or models based on **Extreme Value Theory** (which explain rare events and from there fat tails). Note that all these sophisticated models make it easier to handle the issues of skew or kurtosis. A review of these models and how they can be applied to calculate a **VaR** can be found in Hao Li and Co [59]. To go further into the theory of the bootstrap and particularly into the questions of convergence of bootstrap estimators, one can consult Efron and Tibshirani [40] and Davidson and Hinkley [35].

Non-parametric **VaR** methods have the advantages of being easy to implement as there is no complex step of parameter estimations, and the bootstrap method is an easy to implement simulation process. Non-parametric **VaR** methods use the data and the data only, no prior hypothesis is made on the form of the distribution so, in a certain way, there is no "model risk" when using this method. If there is some skew and kurtosis in the sample, this skew and kurtosis will appear in the simulations as well. Now, of course, for the method to work well, it is necessary for the sample to be large enough and representative enough to give a good grasp of all possible scenarios in the future. An analysis of the performance of the non-parametric **VaR** methods can be found in Barone-Adesi [16].

7.4.6   Example: Boostrap Historical Simulation for a Portfolio VaR

A description of some relevant **VaR** measures prescribed by Basel III is available on the Bank for International Settlements website [14, 15]. In the example here, a 10 business day **VaR**$_{99\%}$ is calculated for a portfolio. The calculation is made by using data from the preceding 12 months, which is the methodology imposed by Basel III. Note that, in the future, this risk measure may be replaced by a 10 business day 97.5% Expected Shortfall, also calibrated on the preceding 12 months' observations.

The investment portfolio π considered here, in the example, consists of five stocks of the DAX index, with an equal weight, for each stock, of 20% at inception. To perform a Bootstrap Historical Simulation of the **VaR** of the portfolio the steps are as follows. First, for the last 12 months, the 255 vectors of the daily prices for the five stocks are extracted. From there, 254 vectors of daily returns are calculated. An empirical law of distribution, for the daily returns of the five stocks, is associated to these 254 vectors of returns, with an equal probability of occurrence of $\frac{1}{254}$ associated to each vector. To simulate a performance over 10 business days, for the five stocks, a Monte Carlo simulation is run. In each simulation, 10 of the 254 vectors of returns are picked, with replacement. For each set of 10 vectors $R(i)$ picked, a return R_{10} over a 10 business day period is calculated by compounding the returns according to the usual formula: $(1 + R_{10}) =$

$(1 + R(1))(1 + R(2)) \cdots (1 + R(10))$. From there, a simulation of the portfolio return is obtained as $\pi' R_{10}$. Therefore, for each simulation of 10 vectors, we end up with a simulation of the portfolio return over a 10 business day period. From these simulated returns, for the portfolio, the quantile corresponding to the lowest 1% returns is calculated, thus giving the **VaR**$_{99\%}$ of the portfolio. In the example here, 10,000 simulations are considered. Of course, the value obtained for the **VaR**$_{99\%}$ of the portfolio may differ slightly from one set of 10,000 simulations to another. The question of an interval of confidence for the **VaR** simulated and of the convergence of the bootstrap method will not be not discussed here, but some elements can be found in Efron and Tibshirani [40].

Finally, note that one of the interesting aspects of the BHS **VaR** is that it preserves, in the simulation, the correlation features (or generally speaking linkage), between the stocks, that may exist in the historical data set. Indeed, if we imagine, for example, that the stock returns all have the same signs, inside each data set vector, this property will of course be preserved in the simulation. So, the property of positive correlation between the stocks (either all going up or all going down every day), also called **cross-sectional correlation** for the data series, will be preserved in the BHS method. What will not be preserved is any possible correlation structure or linkage between consecutive days' returns, as the observations are shuffled randomly to calculate the 10 business days' performance.

The Python code to calculate the BHS **VaR** of this example is given in Listing 7.1

Listing 7.1   Python. BHS **VaR**$_{99\%}$ for a 5 stocks portfolio

```
 1  # First part: data extraction
 2  import pandas as pd
 3  import pandas_datareader.data as web
 4  Tickers = ['DPW.DE', 'ALV.DE', 'BMW.DE', 'BAS.DE', 'FME.DE']
 5  startinput = '2017-01-1'
 6  endinput = '2017-12-31'
 7  S = pd.DataFrame() # creation of the data frame that will contain
        the data set
 8  for t in Tickers:
 9  S[t] = web.DataReader(name = t, data_source='yahoo', start=
        startinput, end= endinput)['Close']
10
11  # Second part: BHS VaR calculation for the portfolio
12  import numpy as np
13  import matplotlib.pyplot as plt
14  # we want to calculate the VaR of the portfolio returns over p
        days. For this, we produce s simulations for the portfolio
        returns over p days. This is done through the bootstrap, by
        choosing with replacement p vectors of stocks returns amongst
        the n calculated.
15  d = len(Tickers) # number of stocks considered
16  p = 10 # number of daily returns considered in each bootstrap
        simulation
17  s = 10000 # number of bootstrap simulations
18  R = pd.DataFrame() # creation of the data frame that will contain
        the five stocks returns
```

```
19  for  t  in  Tickers:  R[t] =S[t]/S[t].shift(1)-1 # calculate the
         stock  returns  from  the  closing  prices  observed
20  R = R[1:]   # eliminate  the  first  value  which  cannot  be  calculated
21  # definition  of  the  weights  for  the  portfolio
22  w = np.zeros(d) # defines  the  shape  of  the  vector  of  allocation
23  for  i  in  range(d):
24      w[i]= 1.0/d # arbitrary  choice  of  constant  weights  for  the
         portfolio  in  this  example
25  # bootstrap  method  used  in  each  of  the  s  simulations
26  n = len(R.index) # number  of  vector  of  returns  from  the  sample
27  perf = np.zeros(s) # array  where  the  portfolio  return  is  saved
         for  each  simulation
28  for  i  in  range(s):
29      idx = np.random.choice(n, p) # select  with  replacement  p
         vectors  of  returns  amongst  n
30      RB = R.iloc[idx]   # extract  the  p  vectors  of  returns  from  the
         return  data  frame
31      RBP = np.prod(RB +1)-1 # calculate  the  returns  over  p  days
         for  each  stock
32      perf[i] = RBP.dot(w) # calculate  the  portfolio  return  over  p
         days
33  plt.hist(perf, 30, histtype='step') # histogram  for  the  portfolio
         return  with  30  buckets
34  plt.grid(True) # grid  added  to  the  histogram
35  print np.percentile(perf, [0, 0.01])
36  print ("The  Var  99%  is:"), np.percentile(perf, [0, 0.01])[1]
```

The result we get from the program for the **VaR**$_{99\%}$ is -6.64%. So, in dollar terms, a one million dollar portfolio is not expected, with 99% chance, to suffer over a 10 business day period a loss of over USD 66,400.

Figure 7.2 represents the histogram for the returns of the portfolio obtained in the simulation.

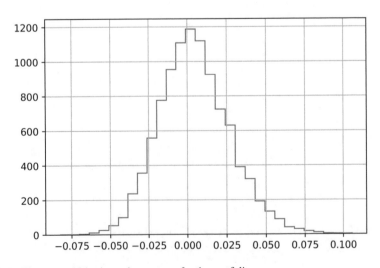

Fig. 7.2 Histogram, 10 business days returns for the portfolio

A Few References

1. Abada, P., Benitob, S., & Lopez, C. (2014). A comprehensive review of value at risk methodologies. *The Spanish Review of Financial Economics, 12*(1), 15–32.
2. Arsac, M., Brie, L., & Genest, B. (2018). *Back-testing of expected shortfall: Main challenges and methodologies.* Chappuis Halder and Co.
3. Artzner, P., Delbaen, F., Eber, J.-M., & Heath, D. (1999). Coherent measures of risk. *Mathematical Finance, 9*, 203–228.
4. Bank for International Settlement. (2016). *Standards: Minimum capital requirements for market risk.* ISBN 978-92-9197-416-0 (online).
5. Bank for International Settlement. (2019). *Minimum capital requirements for market risk.* Jan 2019, Revised Feb 2019, (136 pp.). https://www.bis.org/bcbs/publ/d457.pdf.
6. Bank for International Settlement. (2019). *Instructions for Basel III monitoring.* July 2019. https://www.bis.org/bcbs/qis/biiiimplmoninstr_jul19.pdf.
7. Barone-Adesi, G., & Giannopoulos, K. (2000). Non-parametric VaR Techniques. Myths and Realities. *Economic Notes, 30*, 167–181.
8. Boonen, T. J. (2017). Solvency II solvency capital requirement for life insurance companies based on expected shortfall. *European Actuarial Journal, 7*(2), 405–434.
9. Cai, Z., Fang, Y., & Tian, D. (2018). Econometric modeling of risk measures: A selective review of the recent literature. Working Paper, October 10, 2018.
10. Davidson, A. C., & Hinkley, B. V. (1997). *Bootstrap methods and their application.* Cambridge Series in Statistical and Probabilistic Mathematics (592 pp.). Cambridge: Cambridge University Press.
11. Dowd, K. (2005). *Measuring market risk* (2nd ed., 410 pp.). New York: Wiley.
12. Efron, B., & Tibshirani, R. J. (1994). *An introduction to the bootstrap* (430 pp.). London: Chapman and Hall. New edition, due date December 18, 2019.
13. Engle, R. F., & Manganelli, S. (2001). Value at risk models in finance. ECB Working Paper, No. 75, August 2001.
14. Holden, L. (2008). Some properties of Euler Capital Allocation. Working paper, Norwegian Computing Center, P.O. Box Blindern, NO-314, Oslo.
15. Lie, H., Li, Y., Fan, X., Zhou, Y., Jin, Z., & Liu, Z. (2012). Approaches to VaR, Document Standford University.
16. McNeil, A. J., Frey, R., & Embrechts, P. (2005). *Quantitative risk management: Concepts, techniques and tools.* Princeton: Princeton University Press.
17. Morgan, J. P., & Reuters. (1996). *RiskMetrics^{TM} - Technical Document* (4th ed.). New York, December 17, 1996.
18. Osmundsen, K. K. (2017). Using Expected Shortfall for Credit Risk Regulation, University of Stavanger Working Papers in Economics and Finance 2017/4.
19. Tasche, D. (1999). Risk contribution and performance measurements. Working paper, Technische Universitet Munchen, 1999.
20. Tasche, D. (2008). Capital Allocation to Business Units and Sub-Portfolios: The Euler Principle. arxiv.org/pdf/0708.2542.pdf.

Factor Models

8

In the Security Market Line theorem, the Tangent Portfolio happens to be a single factor, which explains alone all the excess expected returns of all the assets to the risk-free rate, and incidentally explains a portion of their risks, which is called the systematic risk. If now the aim is to explain the risk, i.e. the standard deviation of the returns of all the assets, then the Tangent Portfolio may not be the best instrument to consider, as that is not the specific purpose of this factor. In this chapter, we study techniques to find the best factors to explain the risks, and do not limit ourselves to searching for a single factor. Ideally, the set of factors identified should explain most of the risky assets' variances and correlations, and potentially leave the residual unexplained variations as independent "noises". Two types of factors can be considered: **endogenous factors**, which are statistically derived from the observed variables, i.e. from the observed returns of the assets, and **exogenous factors**, which are explanatory variables added to the model, such as inflation or macro-economic indicators. The normal distribution assumption is maintained here, keeping us in the Markowitz framework. When a factor model follows an additional condition, called the APT condition, it is called an **APT model**. For these models the **fundamental APT theorem** links each factor to a risk premium. In APT models the factors explain all the common sources of risks of the risky assets, which was the primary objective, but also the expected excess returns of the risky assets to the risk-free rate.

8.1 Definitions and Notations

We revisit here the SML equation for risky assets and investment portfolios

$$r^i(t) = r_0 + b^i(r_M(t) - r_0) + \epsilon^i(t)$$

© Springer Nature Switzerland AG 2020
P. Brugière, *Quantitative Portfolio Management*, Springer Texts in Business and Economics, https://doi.org/10.1007/978-3-030-37740-3_8

because:

- in practice the $\epsilon^i(t)$ and $\epsilon^j(t)$ for different assets appear to be correlated and to represent a significant portion of their variances,
- by adding factors in the decomposition of the returns we aim at identifying better the common sources of risks (whatever their remuneration is) and to end up with smaller non-explained residual-specific risks,
- we want to determine the remunerations linked to all identified sources of risks through a non-arbitrage argument,
- we add here the time parameter t to show, from a time series perspective, which parameters are assumed to be fixed and which ones are supposed to vary over time.

Notations

- $R(t) = \begin{pmatrix} r^1(t) \\ \vdots \\ r^d(t) \end{pmatrix}$ the vector of returns for the d risky assets for the period $[t - 1, t]$,

- $A = \begin{pmatrix} a^1 \\ \vdots \\ a^d \end{pmatrix}$ a constant vector and $B = \begin{pmatrix} b^1_1 & \vdots & b^1_K \\ \vdots & \vdots & \vdots \\ b^d_1 & \vdots & b^d_K \end{pmatrix}$ a constant matrix,

- $F(t) = \begin{pmatrix} f^1(t) \\ \vdots \\ f^K(t) \end{pmatrix}$ the vector of K factors calculated at times t and

- $\mathcal{E}(t) = \begin{pmatrix} \epsilon^1(t) \\ \vdots \\ \epsilon^d(t) \end{pmatrix}$ the vector of residual returns for the d risky assets for the period $[t - 1, t]$.

Assumptions We assume that the random vectors $F(t)$ and $\mathcal{E}(t)$ are such that $(F(t), \mathcal{E}(t))$ form an i.i.d. sample of the variable $(F, \mathcal{E})$.

We define $\Sigma_F = \mathbf{Var}(F)$ and $\Sigma_{\mathcal{E}} = \mathbf{Var}(\mathcal{E})$.

Remark 8.1.1 Σ_F is assumed to be invertible because if this was not the case, we could find a vector $u \neq 0$ such that $u'\mathbf{Var}(F)u = 0$, which would imply that $\mathbf{Var}(u'F) = 0$ and $u'F = Cte$. This would imply that some of the factors would be redundant (could be "co-integrated").

Remark 8.1.2 For a sample of size N, for the empirical matrix of variance-covariance of F to be invertible we need $K < N$.

Definition 8.1.1 (K-factor Model)

$$R = A + BF + \mathcal{E} \tag{8.1.1}$$

is a **K-factor model** iff F and $\mathcal{E}$ are variables of $\mathbb{R}^K$ and $\mathbb{R}^d$, with **Var**(F) invertible, $\mathbf{E}(\mathcal{E}) = 0$ and **Cov**$(F, \mathcal{E}) = 0$.

Remark 8.1.3 In the factor model $R = A + BF + \mathcal{E}$, for each risky asset i we have

$$r^i = a^i + \sum_{j=1}^{j=k} b^i_j f^j + \epsilon^i.$$

8.1.1 The Tangent Portfolio as a Factor

In the Markowitz framework the Tangent Portfolio emerges as a natural factor from the SML theorem.

Remark 8.1.4 The SML equation $r^i = r_0 + \beta_T(i)(r_T - r_0) + \epsilon^i$ corresponds to a one-factor model with $a^i = r_0$, $b^i = \beta_T(i)$ and $f^1 = r_T - r_0$.

Remark 8.1.5 In the SML theorem, the Tangent Portfolio is a factor, explaining the excess expected remuneration, but in factor models in general the focus is more on identifying the assets' various sources of risks, which generate their variances and the correlations between their returns.

8.1.2 Endogenous and Exogenous Factors

Remark 8.1.6 Factor models were introduced by Charles Spearman in 1904 in psychometrics.

Remark 8.1.7 In financial econometrics, the factors used are either:

- Macroeconomic factors: ex GDP, inflation rate, unemployment rate... etc, in this case the $F(t)$ are exogenous, i.e. given and observable.
- Fundamental factors: ex market capitalisation, leverage, book/price... etc, which are also exogenous.
- Statistical factors: in this case the $F(t)$ are endogenous factors, and the aim is to determine these $F(t)$. Here, the factors are obtained via some statistical methods

like Principal Component Analysis and they can be interpreted as the returns of some investments and self-financing portfolios which are not correlated.

Remark 8.1.8 Once the factors are estimated or chosen, their variations are regressed over the time series of the observed vector of returns of the assets, in order to calculate the sensibilities (i.e. B).

8.1.3   Standard Form for a Factor Model

In this section we see some alternative ways to write a factor model. We recall the diagonalisation theorem.

Theorem 8.1.1 (Diagonalisation Theorem) *If Σ is symmetric positive definite in $(\mathbb{R}^k, \langle \cdot, \cdot \rangle)$ we can find $v_1, v_2, \cdots, v_k$ in $\mathbb{R}^k$ and $\lambda_1, \lambda_2, \cdots, \lambda_k$ strictly positive such that:*

(1) $\Sigma v_i = \lambda_i v_i$,
(2) $\langle v_i, v_j \rangle = \delta_{i,j}$ (orthonormal basis).

Matricially, if we denote by V the matrix whose vectors columns are the v_i then:

(3) $V'V = VV' = \mathrm{Id}_k$ (orthonormal basis),
(4) $V'\Sigma V = \mathrm{diag}(\lambda_i)$,
(5) $\Sigma = \sum_{i=1}^{i=k} \lambda_i v_i v_i'$.

Remark 8.1.9 The λ_i are the eigenvalues of Σ and the v_i are eigenvectors.

Theorem 8.1.2 (Standard Form (Normalisation of the Factors)) *A K-factor model can be written in the form $R = A + DH + \mathcal{E}$ with $\mathbf{Var}(H) = \mathrm{Id}_K$. This form is called the **standard form** for the factor model.*

Proof Let V be the orthonormal basis of $\mathbf{Var}(F)$ as defined previously.

$$A + BF + \mathcal{E}$$
$$= A + BVV'F + \mathcal{E}$$
$$= A + BV\mathrm{diag}(\sqrt{\lambda_i})\mathrm{diag}(\frac{1}{\sqrt{\lambda_i}})V'F + \mathcal{E}$$
$$= A + \left(BV\mathrm{diag}(\sqrt{\lambda_i})\right)\left(\mathrm{diag}(\frac{1}{\sqrt{\lambda_i}})V'F\right) + \mathcal{E} = A + DH + \mathcal{E}$$

with $D = BV\mathrm{diag}(\sqrt{\lambda_i})$ and $H = \mathrm{diag}(\frac{1}{\sqrt{\lambda_i}})V'F$. Now, $\mathbf{Var}(H) = \mathrm{diag}(\frac{1}{\sqrt{\lambda_i}})\mathbf{Var}(V'F)\mathrm{diag}(\frac{1}{\sqrt{\lambda_i}})$ and $\mathbf{Var}(V'F) = V'\Sigma_F V = \mathrm{diag}(\lambda_i)$. So, $\mathbf{Var}(H) = \mathrm{diag}(\frac{1}{\sqrt{\lambda_i}})\mathrm{diag}(\lambda_i)\mathrm{diag}(\frac{1}{\sqrt{\lambda_i}}) = \mathrm{Id}_K$ and it is easy to verify that $\mathbf{Cov}(F,\mathcal{E}) = 0 \Rightarrow \mathbf{Cov}(H,\mathcal{E}) = 0$. $\qquad\square$

Exercise 8.1.1 Prove Theorem 8.1.1 "intrinsically" by decomposing the vector F in the orthonormal basis $(v_i)_{i\in[\![1,k]\!]}$ of eigenvectors of $\mathbf{Var}(F)$ as

$$R = A + B\left(\sum_{i=1}^{i=k}\langle F, \frac{v_i}{\sqrt{\lambda_i}}\rangle \sqrt{\lambda_i}\, v_i\right) + \mathcal{E}.$$

Corollary 8.1.1 *If the vector of returns R of d risky assets $S_1, \cdots, S_d$ has an invertible matrix of variance-covariance Σ and a matrix of correlation Λ then*

(1) Σ is diagonalisable,
(2) Λ is diagonalisable.

Proof Σ is symmetric positive definite and thus can be diagonalised. Λ is symmetric and also positive definite according to Proposition 3.2.2 and therefore can be diagonalised. $\qquad\square$

8.2   Identifying the Coefficients When the Factors Are Known

We show here that, if the law of (R, F) is known, then the parameters A and B of the factor model (8.1.1) are determined in a unique way. This result leads to a method of estimating A and B from a sample $(R(t), F(t))_{t\in[\![1,T]\!]}$ of (R, F). Indeed, the empirical law defined from the sample enables us, through the plug-in estimator technique, to estimate A and B. We denote by $\mathbf{tr}(\cdot)$ the **Trace**, which is a linear form which associates to a square matrix the sum of its diagonal elements. In the next chapter we will also define the Trace intrinsically.

The following lemma will be useful when doing estimations.

Lemma 8.2.1 (Measure of Dispersion) *Let X be a random variable of $\mathbb{R}^k$, then*

$$\mathbf{E}(\|X\|^2) = \mathbf{tr}(\mathbf{Var}(X)) + \|\mathbf{E}(X)\|^2 \qquad (8.2.1)$$

and $\mathbf{tr}(\mathbf{Var}(X)) = \mathbf{E}(\|X - E(X)\|^2)$.

Proof $\mathbf{Var}(X) = \mathbf{E}(XX') - \mathbf{E}(X)\mathbf{E}(X)'$, so

$$\mathbf{tr}(\mathbf{Var}(X)) = \mathbf{tr}(\mathbf{E}(XX')) - \mathbf{tr}(\mathbf{E}(X)\mathbf{E}(X)')$$
$$= \mathbf{E}(\mathbf{tr}(XX')) - \mathbf{tr}(\mathbf{E}(X)'\mathbf{E}(X))$$

$$= \mathbf{E}(\mathbf{tr}(X'X)) - \|\mathbf{E}(X)\|^2$$

$$= \mathbf{E}(\|X\|^2) - \|\mathbf{E}(X)\|^2,$$

which proves (8.2.1) and the second result follows by replacing X with $X - E(X)$ in Eq. (8.2.1).   □

Proposition 8.2.1 *In a factor model $R = A + BF + \mathcal{E}$ where the law of (R, F) is known, A and B are determined in a unique way by*

$$B = \mathbf{Cov}(R, F)\Sigma_F^{-1} \text{ and } A = \mathbf{E}(R) - B\mathbf{E}(F).$$

A and B are also the solutions to the minimisation problem

$$\min_{A,B} \mathbf{E}\big(\|R - A - BF\|^2\big).$$

Proof $\mathbf{Cov}(R, F) = \mathbf{Cov}(A + BF + \mathcal{E}, F) = \mathbf{Cov}(BF, F) = B\mathbf{Cov}(F, F)$, so $B = \mathbf{Cov}(R, F)\mathbf{Var}(F)^{-1}$. We have $\mathbf{E}(R) = \mathbf{E}(A + BF + \mathcal{E}) = A + B\mathbf{E}(F) \Longrightarrow A = \mathbf{E}(R) - B\mathbf{E}(F)$. For the second result we use that

$$\mathbf{E}\big(\|R - A - BF\|^2\big) = \|\mathbf{E}(R - A - BF)\|^2 + \mathbf{tr}\big(\mathbf{Var}(R - BF)\big),$$

so the problem of optimisation in A and B is reduced to the problem of finding a solution B^* of $\min_{B} \mathbf{tr}\big(\mathbf{Var}(R - BF)\big)$ and for this B^* to take $A^* = \mathbf{E}(R) - B^*\mathbf{E}(F)$. To solve $\min_{B} \mathbf{tr}\big(\mathbf{Var}(R - BF)\big)$ we consider $\phi : \mathbb{R}^{N \times K} \longrightarrow \mathbb{R}$ defined by $\phi(B) = \mathbf{tr}\big(\mathbf{Var}(R - BF)\big)$. The differential of ϕ is defined by

$$d\phi(B)(H) = \mathbf{tr}\big(- 2H\mathbf{Cov}(F, R) + 2H\mathbf{Cov}(F, F)B'\big)$$

$$= -2\mathbf{tr}\big(H(\mathbf{Cov}(F, R) - \mathbf{Cov}(F, F)B')\big)$$

and at the point of extremum this differential has to cancel for every H, which leads to $B = \mathbf{Cov}(R, F)\Sigma_F^{-1}$.   □

We now make some further remarks to show that factor models are particular cases of the Markowitz model when $(F, \mathcal{E})$ is Gaussian.

Remark 8.2.1 If $(F, \mathcal{E})$ is a Gaussian vector then R is a Gaussian vector.

Remark 8.2.2

$$\begin{aligned}
\mathbf{Var}(R) &= \mathbf{Cov}(BF + \mathcal{E}, BF + \mathcal{E}) \\
&= \mathbf{Cov}(BF, BF) + \mathbf{Cov}(\mathcal{E}, \mathcal{E}) \\
&= B\mathbf{Cov}(F, F)B' + \Sigma_{\mathcal{E}} \\
&= B\Sigma_F B' + \Sigma_{\mathcal{E}}.
\end{aligned}$$

So, if $\Sigma_{\mathcal{E}}$ is invertible, the variance-covariance matrix for R is invertible (which is always our hypothesis in a Markowitz framework) as it is the sum of two positive symmetric matrices with one of them strictly positive.

8.2.1 Regression on the Factors

We describe here how to estimate, by linear regression, the coefficients a^i and b^i_k in the factor model $r^i(t) = a^i + \sum_{j=1}^{j=K} b^i_j f^j(t) + \epsilon^i(t)$.

The factors $f^j(t)$ are assumed to be observed as well as the returns $r^i(t)$.

Proposition 8.2.2 *Let $R = A + BF + \mathcal{E}$ be a factor model, for which we observe a sample $(R(t), F(t))$ for $t \in [\![1, T]\!]$. We assume that the empirical matrix of variance-covariance of F, for this sample, denoted $\widehat{\Sigma}_F$, is invertible. Then, to estimate A and B we can use two different approaches which lead to the same results:*

- *Use **plug-in estimators** to calculate, with the empirical probability derived from the sample, $B^* = \widehat{\mathbf{Cov}(R, F)}\widehat{\Sigma}_F^{-1}$ and $A^* = \widehat{\mathbf{E}}(R) - B^*\widehat{\mathbf{E}}(F)$.*
- *Calculate the **ordinary least square estimates** for A and B by solving*

$$\min_{A,B} \sum_{t=1}^{T} \|R(t) - A - BF(t)\|^2.$$

Proof The two methods are equivalent to the two methods described in Proposition 8.2.1 but here in the particular case of a probability which is the empirical probability derived from the sample $\{(R(t), F(t))\}_{t \in [\![1,T]\!]}$. So, according to Proposition 8.2.1 the two results are the same. □

Remark 8.2.3 If $\mathbf{Var}(F)$ is diagonal, from Proposition 8.2.2 we can deduce that $b^i_j = \frac{\mathbf{Cov}(R^i, f^j)}{\mathbf{Var}(f^j)}$, which is the beta of asset i to the factor f^j and that we denote by $\beta_{f_j}(i)$.

8.3 Example of a Factor Model

Consider the three-factor model $R = A + BF + \mathcal{E}$ with three risky assets. We assume that:

- $\mathbf{E}(R) = \begin{pmatrix} 5\% \\ 4\% \\ 6\% \end{pmatrix}$, $B = \begin{pmatrix} 1 & 0 & 0 \\ 0 & 1 & 0 \\ 1 & 0 & 1 \end{pmatrix}$ and $C = \begin{pmatrix} 100\% & 0 & 50\% \\ 0 & 100\% & 0 \\ 50\% & 0 & 100\% \end{pmatrix}$ is the matrix of correlations of the factors.

- $\sigma(f^1) = 15\%$, $\sigma(f^2) = 10\%$, $\sigma(f^3) = 10\%$ are the standard deviations of the factors.

- $(F, \mathcal{E})$ is Gaussian with $\mathbf{Cov}(F, \mathcal{E}) = 0$ and $\mathbf{Var}(\mathcal{E})$ diagonal.

- $\sigma(\epsilon^1) = 5\%$, $\sigma(\epsilon^2) = 5\%$, $\sigma(\epsilon^3) = 5\%$ are the standard deviations of the specific risks represented by $\mathcal{E}$.

- There is a risk-free asset of return $r_0 = 2\%$.

After calculating the law of R and applying Markowitz's results we find that:

(a) The investment portfolio of minimum variance π_a satisfies $\mathbf{E}(R(\pi_a)) = 3.84\%$ and $\sigma(R(\pi_a)) = 10.69\%$.

(b) The Tangent Portfolio of risky allocation π_T satisfies $\pi_T = (0.684, 0.353, -0.037)'$,
$\mathbf{E}(R(\pi_T)) = 4.61\%$ and $\sigma(R(\pi_T)) = 12.74\%$.

(c) The returns of the Tangent Portfolio can be given in terms of the factors as

$$R(\pi_T) = 0.046 + 0.647 f^1 + 0.353 f^2 - 0.037 f^3 + 0.684\epsilon^1 + 0.353\epsilon^2 - 0.037\epsilon^3.$$

(d) From (c) we can derive the β of the three risky assets relative to the Tangent Portfolio and find: $\beta_T(1) = 1.15$, $\beta_T(2) = 0.77$, $\beta_T(3) = 1.53$.

(e) We can verify that for the three assets the SML is satisfied as:

$$5\% = 2\% + 1.15 \times (4.6\% - 2\%),$$

$$4\% = 2\% + 0.77 \times (4.6\% - 2\%),$$

$$6\% = 2\% + 1.53 \times (4.6\% - 2\%).$$

(f) The residuals e^i of the returns of the three risky assets in the SML model expressed as $r^i = r_0 + \beta_T(i)(r_T - r_0) + e^i$ have variance-covariance matrix

$$\begin{pmatrix} 0.004 & -0.007 & 0.001 \\ -0.007 & 0.013 & -0.002 \\ 0.001 & -0.002 & 0.012 \end{pmatrix}$$

and satisfy $\sigma(e^1) = 5.98\%$, $\sigma(e^2) = 11.39\%$, $\sigma(e^3) = 10.92\%$.

As we can see in this example, the decomposition in a three-factor model enables a better explanation of the risks than the decomposition in a (one-factor) SML model because:

- in the three-factor model the residual risks have lower variances than in the SML model,
- in the three-factor model the residual risks are uncorrelated and all common sources of risks have been identified.

This being said, we have not explained here the expected returns of the assets by the risk linked to three factors. This is done in the next section in the context of some particular factor models called APT models.

Remark 8.3.1 (Fama and French Three Factor Model [43]) In this model it is assumed that

$$R_t - r_t^0 1_d = A + b_1(R_t^M - r_t^0) + b_2\text{SMB}_t + b_3\text{HML}_t + \mathcal{E}_t,$$

where

- R_t is the vector of returns of the d risky assets over the period $[t-1, t]$,
- r_t^0 is the risk-free rate (known at time $t-1$) for the period $[t-1, t]$,
- SMB_t is the difference of the returns between the big market capitalisation stocks and the small market capitalisation stocks,
- HML_t is the difference of the returns between the "value stocks" (high book to price ratio) and the "value stocks" (low book to price ratio),
- b_1, b_2, b_3 are fixed vectors (and represent the regression coefficients).

This model was devised in 1993 by Fama and French and still has some strong supporters today. Some extensions were also made by Fama and French in 2014 [45] by adding three extra factors to the model.

8.4 APT Models

The K-factor models presented in this chapter are models for which the conditions of absence of arbitrage opportunities (as defined in Chap. 3) are satisfied, as the matrix of variance-covariance for the returns of the risky assets is invertible. In this section, a second hypothesis for the risky assets is considered, called the **APT condition**, which is a condition of absence of arbitrage opportunities (**AAO** condition) in the reduced model, derived from the factor model, by eliminating the residual risks ϵ^i. When a K-factor model satisfies the APT condition it is called an **APT model**. We derive in this section necessary and sufficient conditions for a K-factor model to be an APT model, and we show that the SML equations define an APT model. We also demonstrate that in an APT model only the risks correlated to the factors are remunerated, which is a result we knew already from the SML theorem, when considering the single factor model $r^i = r_0 + \beta_T(i)(R_T - r_0) + \epsilon^i$. We also show the converse, by proving that if in a factor model only the risk correlated to the factors is remunerated then this factor model is also an APT model.

Definition 8.4.1 (APT Model) A K-factor model $R = A + BF + \mathcal{E}$ is an **APT model** if and only the **AAO** conditions (as defined in Chap. 3) are satisfied in the reduced model $R = A + BF$ (where the "diversifiable" risk $\mathcal{E}$ is neglected).

Theorem 8.4.1 (APT Theorem) *The factor model $R = A + BF + \mathcal{E}$ is an APT model if and only if $\exists \lambda_0, \lambda_1, \lambda_2, \cdots, \lambda_K \in \mathbb{R}$ such that* $\mathbf{E}(R) = \lambda_0 \begin{pmatrix} 1 \\ \vdots \\ 1 \end{pmatrix} +$

$B \begin{pmatrix} \lambda_1 \\ \vdots \\ \lambda_K \end{pmatrix}$, *which we can also write as* $\mathbf{E}(R) = \lambda_0 1_d + B\lambda$ *with* $\lambda_0 \in \mathbb{R}$ *and* $\lambda \in \mathbb{R}^K$.

Remark 8.4.1 If $K < d$ it is possible to build in the reduced model a risk-free portfolio by choosing $\pi \neq 0$ such that $\pi' B = 0$.

- If $\pi' 1_d \neq 0$ we are able to build a risk-free investment portfolio.
- If for all such π, $\pi' 1_d = 0$, we are only able to build risk-free self-financing portfolios.

To demonstrate the APT theorem we start by proving the following two lemmas.

Lemma 8.4.1 *If the APT conditions are satisfied then $\exists \lambda_0$ such that $\forall \pi \in \mathbb{R}^d$, $\pi' B = (0, 0, \cdots, 0) \implies \pi'(A - \lambda_0 1_d) = 0$.*

Proof Let $\pi \in \mathbb{R}^d \setminus \{0\}$ be such that $\pi' B = (0, 0, \cdots, 0)$.

- If $\pi' 1_d \neq 0$ then $\tilde{\pi} = \frac{\pi}{\pi' 1_d}$ is an investment portfolio which is without risk in the reduced model as $R(\tilde{\pi}) = \tilde{\pi}'A + \tilde{\pi}'BF = \tilde{\pi}'A$, which is a constant. Therefore, assuming the **AAO** conditions are met for the reduced model, all such risk-free investment portfolios should have the same return and, denoting this return by λ_0, we should then have $\tilde{\pi}'A = \lambda_0$ and thus $\tilde{\pi}'(A - \lambda_0 1_d) = 0$ and $\pi'(A - \lambda_0 1_d) = 0$.

- If $\pi' 1_d = 0$ then π is a self-financing portfolio without risk, which should therefore satisfy, assuming the **AAO** conditions are met for the reduced model, $\pi'A = 0$ and in this case, for any value of λ_0 considered, $\pi'(A - \lambda_0 1_d) = 0$ as $\pi'A = 0$ and $\pi' 1_d = 0$.   □

Lemma 8.4.2 *The following two assumptions are equivalent:*

(A1) $\forall \pi \in \mathbb{R}^d$, $\pi' B = (0, 0, \cdots, 0) \implies \pi'(A - \lambda_0 1_d) = 0$.
(A2) $\exists \mu \in \mathbb{R}^K$ such that $A - \lambda_0 1_d = B\mu$.

Proof Let $B = (b_1 | \cdots | b_K)$ be the matrix whose columns are made of the vectors b_i and $\text{Vect}\{b_1, b_2, \cdots, b_K\}$ be the vector space generated by $b_1, b_2, \cdots, b_K$ then:

$$(A1) \iff \text{Vect}\{b_1, b_2, \cdots, b_K\}^{\perp} \subset \text{Vect}\{A - \lambda_0 1_d\}^{\perp}$$
$$\iff \text{Vect}\{A - \lambda_0 1_d\} \subset \text{Vect}\{b_1, b_2, \cdots, b_K\}$$
$$\implies A - \lambda_0 1_d \in \text{Vect}\{b_1, b_2, \cdots, b_K\},$$

which proves the lemma. $\qquad\qquad\qquad\qquad\qquad\qquad\qquad\qquad\qquad\qquad\qquad$ □

Proof of the APT Theorem

First implication $\Rightarrow$: According to Lemmas 8.4.1 and 8.4.2 the APT conditions imply that $\exists \lambda_0 \in \mathbb{R}$ and $\mu \in \mathbb{R}^K$ such that $A = \lambda_0 1_d + B\mu$.
Therefore as $\mathbf{E}(R) = A + B\mathbf{E}(F)$ we get $\mathbf{E}(R) = \lambda_0 1_d + B\mu + B\mathbf{E}(F) = \lambda_0 1_d + B\lambda$ if we define λ by $\lambda = \mu + \mathbf{E}(F)$.
Second implication $\Leftarrow$: We assume now that $\exists \lambda_0 \in \mathbb{R}$ and $\lambda \in \mathbb{R}^K$ such that $\mathbf{E}(R) = \lambda_0 1_d + B\lambda$ and want to prove that the APT conditions are satisfied.

Let π be a portfolio without risk in the reduced model

- π investment portfolio without risk in the reduced model $\implies \pi' B = 0$ and in this case, $\mathbf{E}(R(\pi)) = \pi' \mathbf{E}(R) = \lambda_0 \pi' 1_d = \lambda_0$. So, the first **AAO** condition that all risk-free investment portfolios have the same return is satisfied.
- π self-financing portfolio without risk in the reduced model $\implies \pi' B = 0$ and in this case $\mathbf{E}(R(\pi)) = \pi' \mathbf{E}(R) = \lambda_0 \pi' 1_d = 0$. So, the second **AAO** condition that all risk-free self-financing portfolios have a return of zero is satisfied. So all APT conditions are satisfied and this finishes the proof of the APT theorem. □

Remark 8.4.2 If there is no possibility to build risk-free investment portfolios in the reduced economy then in the APT model any value can be taken for λ_0. In this case 1_d is a linear combination of the vector columns of B, as will be demonstrated in the next section, and if there is a risk-free asset in the economy of return r_0 we will take $\lambda_0 = r_0$.

Remark 8.4.3 If there is a risk-free asset in the economy of return r_0 and a possibility to build risk-free investment portfolios in the reduced economy then in general it is expected that $r_0 = \lambda_0$.

Remark 8.4.4 λ_k is called the remuneration of the risk in excess of λ_0 linked to factor k. This definition is particularly meaningful when there is a risk-free rate r_0 in the economy and that $r_0 = \lambda_0$.

8.4.1   Example of an APT Model

We consider the two factor model $R = A + BF + \mathcal{E}$ with three risky assets satisfying:

$$\mathbf{E}(R) = \begin{pmatrix} 5\% \\ 8\% \\ 5\% \end{pmatrix}, B = \begin{pmatrix} 1 & 0 \\ 0 & 1 \\ \frac{1}{3} & \frac{1}{3} \end{pmatrix}, \mathbf{E}(\mathcal{E}) = 0, \mathbf{Cov}(F, \mathcal{E}) = 0 \text{ and } \mathbf{Var}(F) \text{ invertible.}$$

According to the APT theorem, this factor model (on risky assets) is an APT model iff we can find $\lambda_0, \lambda_1, \lambda_2$ such that

$$\begin{cases} 5\% = \lambda_0 + \lambda_1 \\ 8\% = \lambda_0 + \lambda_2 \\ 5\% = \lambda_0 + \frac{1}{3}\lambda_1 + \frac{1}{3}\lambda_2. \end{cases}$$

Here, $\lambda_0 = 2\%$, $\lambda_1 = 3\%$, $\lambda_2 = 6\%$ solve the system, so we can conclude that the APT conditions are satisfied.

According to the APT theorem, any risk free-investment portfolio in the reduced model should have a return of λ_0. If we consider $\pi = (-1, -1, 3)'$ then π is an investment portfolio without risk in the reduced model, as the contributions of the two factors in this portfolio cancel out. Now, when we calculate the expected return of this portfolio we find $-1 \times 5\% + (-1) \times 8\% + 3 \times 5\% = 2\%$, which is equal to $\lambda_0 = 2\%$, as expected.

Remark 8.4.5 (Entering a Consistent Vector M of Expected Returns in a Model) In practice, an APT approach can be used to choose the vector M of expected returns to be used by a portfolio manager in a Markowitz optimisation problem. Assume, for example, that the universe of investment is made of the 500 stocks of the S&P500. Then, instead of entering in the model 500 independent predictions of expected returns, which may lead to some unwanted inconsistencies, one may regress the historical returns of the stocks on a limited number of factors and then calculate an expected return for each stock by using predictions on each factor and the regression coefficients. Then, if there is a specific view on a particular stock a top-up return can be added. By doing so, the input of M is probably more consistent and simple than if the choice was made to enter 500 independent individual predictions (unless the choice is made to simply use historical returns as predictions for the future).

8.4.2   Further Remarks

Proposition 8.4.1 *The following two assumptions are equivalent:*

(A1) *There is no risk-free investment portfolio in the reduced model* $R = A + BF$.
(A2) $\text{Vect}\{1_d\} \subset \text{Vect}\{b_1, b_2, \cdots, b_K\}$.

Proof

$$(A1) \iff (x'B = 0 \implies x'1_d = 0)$$
$$\iff \text{Vect}\{b_1, b_2, \cdots, b_K\}^{\perp} \subset \text{Vect}\{1_d\}^{\perp}$$
$$\iff \text{Vect}\{1_d\} \subset \text{Vect}\{b_1, b_2, \cdots, b_K\}. \qquad \square$$

Remark 8.4.6 Proposition 8.4.1 implies that when the APT conditions are satisfied with no possibility to build a risk-free investment portfolio in the reduced economy with the risky assets, then λ_0 is not determined in a unique way in the decomposition $\mathbf{E}(R) = \lambda_0 1_d + B\lambda$ as the vectors $1_d, b_1, \cdots, b_K$ are not independent. In this case, we can take any value for λ_0 (and not only the value zero).

8.4.3   Standard Form for an APT Model

Proposition 8.4.2 (Standard Form of an APT Model) *A factor model is an APT model iff it can be written in the standard form*

$$R = \lambda_0 1_d + BG + \mathcal{E},$$

where $\mathbf{Var}(G)$ *is invertible,* $\mathbf{E}(\mathcal{E}) = 0$ *and* $\mathbf{Cov}(G, \mathcal{E}) = 0$.

Proof We consider a factor model $R = A + BF + \mathcal{E}$. If the APT conditions are satisfied then $A = \lambda_0 1_d + B\lambda$ and $R = \lambda_0 1_d + B\lambda + BF + \mathcal{E} = \lambda_0 1_d + B(F + \lambda) + \mathcal{E}$, which proves the result with $G = F + \lambda$. Conversely, if a factor model is of the form $R = \lambda_0 1_d + BG + \mathcal{E}$ then $\mathbf{E}(R) = \lambda_0 1_d + B\lambda$ with $\lambda = \mathbf{E}(G)$, which implies, according to the APT theorem, that it is an APT model. $\qquad \square$

8.5   Alternative Definition of an APT Model

We prove in this section that the APT conditions are satisfied in a K-factor model if and only if **only the risk correlated with the factors is remunerated**. The Security Market Line equation, which can be seen as a particular way to write a Markowitz model as a one-factor model, illustrates this property and indeed satisfies the APT conditions. There are several ways to write a Markowitz model as a factor model (by decomposing the variance-covariance matrix in different ways) but in any decomposition satisfying the APT conditions only the risk correlated with the factors is remunerated.

Proposition 8.5.1 (Alternative Definition of an APT Model) *Let $R = A + BF + \mathcal{E}$ be a K-factor model, then the following assumptions are equivalent:*

(A1) *The APT conditions are satisfied.*
(A2) $\exists \lambda_0$ *such that $\forall \pi$ portfolios* $\mathbf{Cov}(R(\pi), F) = 0 \Longrightarrow \mathbf{E}(R(\pi)) = \lambda_0 \pi' 1_d$.

Proof Let's assume (A2). It is easy to see that (A2) $\Longleftrightarrow$ $(\tilde{A}2)$, where

$$(\tilde{A}2):\ \exists \lambda_0,\ \forall x \in \mathbb{R}^d,\ x'B\Sigma_F = 0 \Longrightarrow x'\mathbf{E}(R) = \lambda_0 x' 1_d$$

$$\Longleftrightarrow \exists \lambda_0 \in \mathbb{R},\ \forall x \in \mathbb{R}^d,\ x'B = 0 \Longrightarrow x'(\mathbf{E}(R) - \lambda_0 1_d) = 0$$

$$\Longleftrightarrow \exists \lambda_0 \in \mathbb{R},\ \text{Vect}\{b_1, b_2, \cdots b_K\}^{\perp} \subset \text{Vect}\{\mathbf{E}(R) - \lambda_0 1_d\}^{\perp}$$

$$\Longleftrightarrow \exists \lambda_0 \in \mathbb{R},\ \text{Vect}\{(\mathbf{E}(R) - \lambda_0 1_d\} \subset \text{Vect}\{b_1, b_2, \cdots b_K\}$$

$$\Longleftrightarrow \exists \lambda_0 \in \mathbb{R},\ \lambda \in \mathbb{R}^K,\ \mathbf{E}(R) - \lambda_0 1_d = B\lambda$$

$$\Longleftrightarrow \text{(A1) according to the APT theorem.} \qquad \square$$

Remark 8.5.1 (A2) is equivalent to saying that any self-financing portfolio with zero correlation with the factors has an expected return of zero and that all investment portfolios with zero correlation with the factors have the same expected return.

Remark 8.5.2 In an APT model the initial purpose of explaining the correlated risks is explained but on top of that the explanation of all risk premia and excess returns to the risk free rate is obtained as well.

8.5.1 Estimation of the Risk Premia in an APT Model

If $R = A + BF + \mathcal{E}$ is an APT model with given observable factors F, then once the estimators A^* and B^* are obtained by minimising $\hat{\mathbf{E}}(\|R - A - BF\|^2)$, λ_0^* and λ^* can be found by minimising $\|A^* + B^*\hat{\mathbf{E}}(F) - \lambda_0 1_d - B^*\lambda\|$. Here $\hat{\mathbf{E}}$ designates the expectation calculated with the empirical probability on (R, F) derived from the sample.

In this chapter we assumed that the factors were given observable variables and indicated some ways to estimate the fixed parameters of the model. In the next chapter we see how some factors can be identified statistically from the variance-covariance matrix of the returns, with the fixed parameters being estimated simultaneously.

A Few References

1. Ang, N. (2014). *Asset management: A systematic approach to factor investing.* Oxford: Oxford University Press.
2. Bai, J., & Ng, S. (2002, January). Determining the number of factors in approximate factor models. *Econometrica, 70*(1), 191–221.
3. Chamberlain, G. (1983). Funds, factors and diversification in arbitrage pricing models. *Econometrica, 51*, 1305–1323.
4. Chen, N., & Ingersoll, J. (1983). Exact pricing in linear factor models with infinitely many assets: A note. *Journal of Finance, 38*, 985–988.
5. Dybvig, P., & Ross, S. (1985). Yes, the APT is testable. *Journal of Finance, 40*, 1173–1188.
6. Fama, E., & French, K. (1993). Common risk factors in the returns on stocks and bonds. *Journal of Financial Economics, 33*, 3–56.
7. Fama, E. F., & French, K. R. (2015). A five-factor asset pricing model. *Journal of Financial Economics, 116*(1), 1–22.
8. Jobson, J. (1982). A multivariate linear regression test of the arbitrage pricing theory. *Journal of Finance, 37*, 1037–1042.
9. Ross, S. A. (1976). The arbitrage theory of capital asset pricing. *Journal of Economic Theory, 13*(3) (December), 341–360.
10. Roll, R., & Ross, S. (1980). An empirical investigation of the arbitrage pricing theory. *Journal of Finance, 35*, 1073–1103.
11. Sheikh, A. (1996). *BARRA's Risk Models.* Copyright ©1996 BARRA, Inc.
12. Wang, J., & Zivot, E. (2005, December). *Modeling financial time series with S-PLUS* (2nd ed.). New York: Springer.

Identification of the Factors

9

In this chapter we use Principal Component Analysis to study the returns of a set of risky assets and to identify the most relevant factors explaining their variations. The residual risks will be uncorrelated with the factors by construction, and therefore the general conditions of a factor model are satisfied. The number of factors chosen will be based on the percentage of the total variance they explain. If a large portion of the total variance is explained then for most stocks the residual risks will be small.

9.1 Total Inertia and Trace of the Variance-Covariance Matrix

We denote by $\mathbf{tr}(\cdot)$ the Trace operator, which associates to a square matrix the sum of its diagonal components. This operator can be defined intrinsically by considering for any linear transformation f of $\mathbb{R}^d$ and any othonormal basis $(e_i)_{i \in [\![1,d]\!]}$ of $\mathbb{R}^d$ the quantity $\mathbf{tr}(f) = \sum_{i=1}^{d} \langle f(e_i), e_i \rangle$. We show in this section that applying the trace operator to the variance-covariance matrix defines a natural measure of the dispersion (or inertia) of the related random variable. The aim of principal component analysis is to represent (project) a random variable of $\mathbb{R}^d$ in a lower dimensional space, while keeping most of the dispersion (inertia) of the variable. So, principal component analysis is about compressing the data (by linear projection) while trying to lose as little information as possible.

Property 9.1.1 Let $x = (x^1, \cdots, x^d)'$ and $y = (y^1, \cdots, y^d)'$ be two vectors of $\mathbb{R}^d$, then $x'y = \mathbf{tr}(xy')$.

Proof By the properties of the trace, $\mathbf{tr}(xy') = \mathbf{tr}(y'x) = y'x$ and $y'x = x'y$, therefore $\mathbf{tr}(xy') = x'y$. $\qquad\square$

Definition 9.1.1 (Dispersion) $\mathbf{E}\left(\|Z - \mathbf{E}(Z)\|^2\right)$ is called the **dispersion** of Z.

© Springer Nature Switzerland AG 2020
P. Brugière, *Quantitative Portfolio Management*, Springer Texts in Business and Economics, https://doi.org/10.1007/978-3-030-37740-3_9

Remark 9.1.1

- In dimension 1 the definition of the dispersion corresponds to the usual definition of the variance.
- As proved in Lemma 8.2.1, $\mathbf{E}\left(\|Z - \mathbf{E}(Z)\|^2\right) = \mathbf{tr}\left(\mathbf{Var}(Z)\right)$.
- $\mathbf{tr}\left(\mathbf{Var}(Z)\right) = \sum\limits_{i=1}^{i=d} \lambda_i$, where the λ_i are the eigenvalues of $\mathbf{Var}(Z)$.

Definition 9.1.2 (Inertia) The **inertia** of the sample $z_1, \cdots z_n$ of the random variable Z is the quantity

$$\frac{1}{n} \sum_{i=1}^{n} \|z_i - \bar{z}\|^2,$$

where $\bar{z} = \frac{1}{n} \sum\limits_{i=1}^{n} z_i$.

Remark 9.1.2 The inertia of a sample of Z is the dispersion calculated for the empirical probability derived from the sample.

9.2   Total Inertia of the Projection

We show in this section that to maximise the dispersion of the projection of a random vector, it has to be projected on the space containing the eigenvectors of the highest eigenvalues of its variance-covariance matrix.

Proposition 9.2.1 *Let Z be a random variable in $\mathbb{R}^d$, $\mathbf{Var}(Z)$ its matrix of variance-covariance and $\lambda_1 \geq \lambda_2 \geq \cdots \geq \lambda_n \geq 0$ the eigenvalues of $\mathbf{Var}(Z)$. For any $k \leq d$, let $\mathcal{B}_k$ be the set of all orthonormal families of k vectors $(x_i)_{i \in [\![1,k]\!]}$ of $\mathbb{R}^d$, then for any $k \in [\![1, d]\!]$*

$$\max_{(x_i)_{i \in [\![1,k]\!]} \in \mathcal{B}_k} \sum_{i=1}^{i=k} x_i' \mathbf{Var}(Z) x_i = \sum_{i=1}^{i=k} \lambda_i.$$

Proof We solve the maximization problem

$$(P) \begin{cases} \sup \sum\limits_{i=1}^{i=k} x_i' \mathbf{Var}(Z) x_i \\ x_i' x_j = \delta_{i,j}. \end{cases}$$

The Lagrangian is $\sum_{i=1}^{i=k} x_i' \mathbf{Var}(Z) x_i - \sum_{i\neq j} \lambda_{i,j} x_i' x_j - \sum_{i=1}^{i=k} \lambda_{i,i}(x_i' x_i - 1)$. We have

$$\frac{\partial L}{\partial x_i} = 0 \iff 2x_i' \mathbf{Var}(Z) - 2\lambda_{i,i} x_i' - \sum_{j\neq i} \lambda_{i,j} x_j' = 0$$

$$\iff 2\mathbf{Var}(Z)x_i - 2\lambda_{i,i} x_i - \sum_{j\neq i} \lambda_{i,j} x_j = 0.$$

As a consequence $\mathbf{Var}(Z)V \subset V$ with $V = vect\{x_1, \cdots, x_k\}$ and if we note $\mathbf{Var}(Z)_{|V}$ the restriction of $\mathbf{Var}(Z)$ to V we have:

(a) $\mathbf{Var}(Z)_{|V}$ is symmetric definite positive
(b) the eigenvalues of $\mathbf{Var}(Z)_{|V}$ are eigenvalues of $\mathbf{Var}(Z)$
(c) $\sum_{i=1}^{i=k} x_i' \mathbf{Var}(Z) x_i = \sum_{i=1}^{i=k} x_i' \mathbf{Var}(Z)_{|V} x_i = \mathrm{Trace}(\sum_{i=1}^{i=k} \mathbf{Var}(Z)_{|V} x_i x_i') = \mathrm{Trace}(\mathbf{Var}(Z)_{|V} Id_{|V}) = \mathrm{Trace}(\mathbf{Var}(Z)_{|V})$ and $\mathrm{Trace}(\mathbf{Var}(Z)_{|V})$ is the sum of the k eigenvalues of $\mathbf{Var}(Z)_{|V}$, which according to (b) is the sum of k eigenvalues of $\mathbf{Var}(Z)$ and the sum is maximal if the eigenvalues are the k largest ones and in this case the sum equals $\sum_{i=1}^{i=k} \lambda_i$. $\qquad\square$

Corollary 9.2.1 *Let H_k denote a vector subspace of $\mathbb{R}^d$ of dimension k (with $k \leq d$). Let $\mathcal{H}_k$ be the set of all vector spaces H_k. Let p_{H_k} be the orthogonal projection on H_k, then*

$$\sup_{H_k \in \mathcal{H}_k} \mathbf{E}\big(\|p_{H_k}(Z - \mathbf{E}(Z))\|^2\big) = \sum_{i=1}^{i=k} \lambda_i.$$

Proof Let $(x_i)_{i \in [\![1,k]\!]}$ be an orthonormal basis of H_k. Then

$$p_{H_k}(Z - \mathbf{E}(Z)) = \sum_{i=1}^{i=k} x_i'(Z - \mathbf{E}(Z)) x_i \text{ and}$$

$$\|p_{H_k}(Z - \mathbf{E}(Z))\|^2 = \sum_{i=1}^{i=k} (x_i'(Z - \mathbf{E}(Z)))^2 = \sum_{i=1}^{i=k} x_i'(Z - \mathbf{E}(Z))(Z - \mathbf{E}(Z))' x_i.$$

So, $\mathbf{E}\big(\|p_{H_k}(Z - \mathbf{E}(Z))\|^2\big) = \sum_{i=1}^{i=k} x_i' \mathbf{E}\big((Z - \mathbf{E}(Z))(Z - \mathbf{E}(Z))'\big) x_i = \sum_{i=1}^{i=k} x_i' \mathbf{Var}(Z) x_i$.
So, the result follows from Proposition 9.2.1. $\qquad\square$

Exercise 9.2.1 Show that $\inf_{H_k \in \mathcal{H}_k} \mathbf{E}\big(\|p_{H_k}(Z - \mathbf{E}(Z))\|^2\big) = \sum_{i=d-k+1}^{i=d} \lambda_i$.

9.3   Principal Component Analysis and Factors

We consider d risky assets with vector of returns R, matrix of variance-covariance Σ and matrix of correlations Λ. We assume that Σ is invertible. We can conduct a **Principal Component Analysis** on Σ or on the matrix of correlations Λ. In the first case this leads to a factor analysis for R with (1) an approximation of the returns of the d assets by a chosen number of factors and (2) a visualisation of the betas of the assets to these factors. In the second case, the analysis is done on the returns of the risky assets renormalised by their standard deviations.

9.3.1   PCA of the Matrix of Variance-Covariance

Let $(b_i)_{i \in [\![1,d]\!]}$ be an orthonormal basis of $(\mathbb{R}^d, \langle \cdot, \cdot \rangle)$ of eigenvectors of $\mathbf{Var}(R)$ corresponding to the eigenvalues $\lambda_1 \geq \lambda_2 \cdots \geq \lambda_d > 0$. We can write

$$R = \mathbf{E}(R) + \sum_{i=1}^{k} \langle b_i, R - \mathbf{E}(R) \rangle b_i + \mathcal{E} \text{ with } \mathcal{E} = \sum_{i=k+1}^{d} \langle b_i, R - \mathbf{E}(R) \rangle b_i. \quad (9.3.1)$$

Let $B = (b_1 | \cdots | b_k)$ and $F = (f^1, \cdots, f^k)'$ with $f^i = \langle b_i, R - \mathbf{E}(R) \rangle$. Then Eq. (9.3.1) can be expressed as

$$R = \mathbf{E}(R) + BF + \mathcal{E}. \quad (9.3.2)$$

Proposition 9.3.1 *In Eq. (9.3.2) we have* $\mathbf{Var}(F) = \text{diag}(\lambda_i)$ *for* $i \in [\![1, k]\!]$, $\mathbf{E}(\mathcal{E}) = 0$ *and* $\mathbf{Cov}(F, \mathcal{E}) = 0$.

Proof Let's demonstrate the three points:

- $\mathbf{Cov}(f^i, f^j) = \mathbf{Cov}\big(b_i'(R - \mathbf{E}(R)), b_j'(R - \mathbf{E}(R))\big) = b_i'\mathbf{Var}(R)b_j = \delta_{i,j}\lambda_i$. So, $\mathbf{Var}(F) = \text{diag}(\lambda_i)$ for $i \in [\![1, k]\!]$.

- $\mathbf{E}(\mathcal{E}) = \sum_{i=k+1}^{d} \langle b_i, \mathbf{E}(R - \mathbf{E}(R)) \rangle b_i = 0$.

- $\mathbf{Cov}(f^i, \mathcal{E}) = \mathbf{Cov}\left(b_i'(R - \mathbf{E}(R)), \sum_{j=k+1}^{d} \langle b_j, R - \mathbf{E}(R) \rangle b_j \right)$

$$= \sum_{j=k+1}^{d} b_i'\mathbf{Cov}\big(R - \mathbf{E}(R), R - \mathbf{E}(R)\big)b_j b_j'$$

$$= \sum_{j=k+1}^{d} b_i' \mathbf{Var}(R) b_j b_j'$$

$$= 0$$

as $i \in [\![1, k]\!]$ while $j \in [\![k + 1, d]\!]$ and $\forall i \neq j$, $b_i' \mathbf{Var}(R) b_j = 0$.

Therefore, $\forall i \in [\![1, k]\!]$, $\mathbf{Cov}(f^i, \mathcal{E}) = 0$ and consequently $\mathbf{Cov}(F, \mathcal{E}) = 0$. $\qquad\square$

Corollary 9.3.1 *The decomposition of the returns according to the PCA decomposition as expressed in Eq. (9.3.2) defines a factor model for the returns of the risky assets.*

Proof Immediate according to Proposition 9.3.1 and the fact that $\mathbf{Var}(F) = \text{diag}(\lambda_i)$ implies that $\mathbf{Var}(F)$ is invertible. $\qquad\square$

Remark 9.3.1 (Visualisation of the n Returns in a Two-Dimensional PCA)

(1) The n vectors of returns of the d assets are represented by the n points $(f^1(t_i), f^2(t_i))_{i \in [\![1,n]\!]}$. The information is compressed from dimension d to dimension 2. The d assets are "summarised" by the two factors f^1 and f^2, which are called the **principal components**.

(2) $\mathbf{tr}(\mathbf{Var}(b_1 f^1 + b_2 f^2)) = \mathbf{Var}(f^1) + \mathbf{Var}(f^2) = \lambda_1 + \lambda_2$ and therefore the proportion of the dispersion explained by the two-dimensional PCA is the ratio of $\lambda_1 + \lambda_2$ by the sum of the eigenvalues λ_i.

Remark 9.3.2 (Visualisation of the Beta to the Factors in a Two-Dimensional PCA)

(1) For each asset i, its random return R^i is summarised by the quantity $b_1^i f^1 + b_2^i f^2$ which represents, according to Proposition 8.2.1, the best approximation of R^i (from a variance perspective) by a linear combination of f^1 and f^2. In the PCA we can visualise the $(b_1^i, b_2^i)_{i \in [\![1,d]\!]}$ for the d assets.

(2) As $\mathbf{Var}(F)$ is diagonal, as noted in Remark 8.2.3, the $(b_1^i, b_2^i)_{i \in [\![1,d]\!]}$ are the beta of the assets to f^1 and f^2 i.e., $b_j^i = \frac{\mathbf{Cov}(R^i, f^j)}{\mathbf{Var}(f^j)} = \beta_{f_j}(i)$. This can also be proved by direct calculation of $\mathbf{Cov}(R^i, f^j)$.

Exercise 9.3.1 Let S be the space of random variables of $\mathbb{R}^d$ with the scalar product $\langle X, Y \rangle = \mathbf{E}(X'Y)$. Show that $b_1 f^1 + b_2 f^2$ is the orthogonal projection of $R - \mathbf{E}(R)$ on the space of random variables $\{f^1 v + f^2 w / v, w \in \mathbb{R}^d\}$.

Remark 9.3.3 When doing a PCA based on observations, all the quantities considered are calculated with the probability derived from the sample (the empirical probability) and not from the true (unknown) probability of the model.

Proposition 9.3.2 (Betas Circle) *Let $\beta_{fj}(i)$ be the beta of asset i relative to the factor j. Then*

$$\sum_{j=1}^{d} \left(\beta_{fj}(i)\right)^2 = 1.$$

Proof Let $(e_i)_{i\in[\![1,d]\!]}$ be the canonical orthonormal basis of $(\mathbb{R}^d, \langle \cdot, \cdot \rangle)$. We have $\beta_{fj}(i) = b_j^i = e_i' b_j$. Now, as $(b_j)_{j\in[\![1,d]\!]}$ is an orthonormal basis of $\mathbb{R}^d$ we have

$$\sum_{j=1}^{d} \left(e_i' b_j\right)^2 = \|e_i\|^2 = 1.$$

$\square$

Corollary 9.3.2 *The points $(b_1^i, b_2^i)_{i\in[\![1,d]\!]}$ for the d assets are all within a circle of radius 1.*

Proof Direct consequence of Proposition 9.3.2. $\square$

Proposition 9.3.3 (Correlations Circle) *Let $\rho(R_i, f_j)$ be the correlation between the return of asset i and the factor j. Then*

$$\sum_{j=1}^{d} \rho^2(R_i, f_j) = 1.$$

Proof $\rho(R_i, f_j) = \frac{\mathbf{Cov}(R_i, f_j)}{\sigma(R_i)\sigma(f_j)}$ with $\mathbf{Cov}(R_i, f_j) = \lambda_j \langle e_i, b_j \rangle$ and $\sigma(f_j) = \sqrt{\lambda_j}$. So,

$$\rho(R_i, f_j) = \frac{\sqrt{\lambda_j}\langle e_i, b_j \rangle}{\sigma(R_i)}$$

and

$$\sum_{j=1}^{d} \rho^2(R_i, f_j) = \frac{1}{\sigma^2(R_i)} \sum_{j=1}^{d} \lambda_j (e_i' b_j)^2.$$

Now,

$$\sigma^2(R_i) = \mathbf{Var}(e_i' R)$$

$$= \mathbf{Var}(e_i' \sum_{j=1}^{d} b_j f_j)$$

$$= \sum_{j=1}^{d} (e_i' b_j)^2 \mathbf{Var}(f_j)$$

$$= \sum_{j=1}^{d} (e_i' b_j)^2 \lambda_j,$$

which proves the result. $\qquad\qquad\qquad\qquad\qquad\qquad\qquad\qquad\qquad\qquad\square$

9.3.2 PCA of the Correlation Matrix

Let $\tilde{R}$ be the vector of returns whose ith-component is defined by $\tilde{R}^i = \frac{R^i}{\sigma(R^i)}$. $\tilde{R}$ is the vector of the returns of the assets, after renormalisation of each return by its standard deviation. Therefore, the matrix of variance-covariance of $\tilde{R}$ is the matrix of correlation of R. When the focus is more on explaining the correlations than the variance, the PCA is applied to the matrix Λ.

9.4 Principal Components and Eigenvalues Visualisation

In this example, we see how we can visualise the matrix of variance-covariance between two risky assets, its eigenvalues and its eigenvectors, from the observations of the returns of these two assets.

Example 9.4.1 We consider an economy with two risky assets whose returns r_1 and r_2 follow the one-factor model described by

$$\begin{pmatrix} r_1 \\ r_2 \end{pmatrix} = \begin{pmatrix} 1 \\ 1.2 \end{pmatrix} f + \begin{pmatrix} 0.6 & 0 \\ 0 & 0.6 \end{pmatrix} \begin{pmatrix} e_1 \\ e_2 \end{pmatrix},$$

with f, e_1 and e_2 independent of variance 1 and expectation zero.

In Fig. 9.1 we plot 800 simulations of $\begin{pmatrix} r_1 \\ r_2 \end{pmatrix}$.

We have the following results:

- the eigenvalues of the variance-covariance matrix are 2.80 and 0.36 in the model and 3.06 and 0.57 from the sample,
- the measure of dispersion is 3.16 in the model and 3.63 for the sample,
- $\begin{pmatrix} 0.64 \\ 0.77 \end{pmatrix}$ and $\begin{pmatrix} 0.77 \\ -0.64 \end{pmatrix}$ are eigenvectors of norm 1 in the model,
- $\begin{pmatrix} 0.70 \\ 0.71 \end{pmatrix}$ and $\begin{pmatrix} 0.71 \\ -0.70 \end{pmatrix}$ are eigenvectors of norm 1 derived from the empirical variance-covariance matrix of the sample,

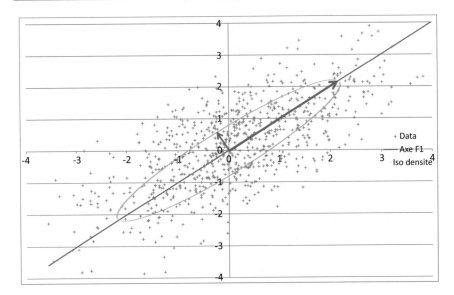

Fig. 9.1 In red, the axis corresponding to the eigenvector of maximum eigenvalue

- in the graph, the red axis corresponds to the sample eigenvector of the highest eigenvalue. It is on this axis that the projection of the points from the sample exhibit the greater dispersion,
- the dispersion of the projections of the sample points on the red axis is 3.06, which is the largest eigenvalue for the sample variance-covariance matrix,
- the green ellipse on the chart represents points for which the density function of a normal law having the mean of the sample and the variance-covariance matrix of the sample is constant,
- the lengths of the axes (in purple) of the ellipse are proportional to the eigenvalues of the sample variance-covariance matrix.

9.5   Python: Application to the DAX 30 Components

We see in this section how a PCA can be executed for the stocks composing the DAX 30 and from there, using the first two components of the PCA, how a two-factor analysis for the returns of the stocks can be conducted.

A principal component analysis can be conducted either by diagonalising the 30×30 matrix of variance-covariance for the daily returns of the 30 stocks or by doing a **Singular Value Decomposition (SVD)** of the $30 \times n$ matrix of the daily centred returns (i.e. for each stock the average return on all observation dates is subtracted from each of the daily returns observed for that stock). If dataRC is the matrix of the daily centred returns for the 30 components of the DAX Index, then each of the 30 rows of dataRC represents the returns of a particular stock over the n periods and according to the Singular Value Decomposition theorem it is possible to decompose dataRC as

$$\text{dataRC} = \begin{pmatrix} b_1| & \cdots & |b_{30} \end{pmatrix} \begin{pmatrix} \dfrac{\text{diag}(\sigma_i)}{0} \end{pmatrix} \begin{pmatrix} v_1^T \\ \cdots \\ v_n^T \end{pmatrix} \tag{9.5.1}$$

where the column vectors v_i form an orthonormal basis of $\mathbb{R}^n$, the vectors b_i form an orthonormal basis of $\mathbb{R}^{30}$ and the row vectors v_i^T denotes the transpose of the column vectors v_i. According to this decomposition we get,

$$\frac{1}{n}\text{dataRC.dataRC}^T = \begin{pmatrix} b_1| & \cdots & |b_{30} \end{pmatrix} \left(\text{diag}(\tfrac{1}{n}\sigma_i^2)\right) \begin{pmatrix} b_1^T \\ \vdots \\ b_{30}^T \end{pmatrix} \tag{9.5.2}$$

and the b_i are eigenvectors of the 30×30 matrix of empirical variance-covariance $\frac{1}{n}\text{dataRC}^T.\text{dataRC}$ of eigenvalues $\frac{1}{n}\sigma_i^2$.

In Python with scikit-learn the v_i, b_i and σ_i are calculated through an SVD decomposition of dataRC.

9.5.1 Factors Explaining the Variance for the DAX 30 Components

The n observed vectors of centred returns $\text{RC}(t_i) = (\text{RC}_1(t_i), \cdots, \text{RC}_{30}(t_i))'$ are projected into the two-dimensional subspace of $\mathbb{R}^{30}$ generated by the eigenvectors b_1 and b_2 with the two highest eigenvalues. The vectors represented in the PCA are therefore the n vectors $\langle \text{RC}(t_i), b_1 \rangle b_1 + \langle \text{RC}(t_i), b_2 \rangle b_2$.

The Python code given in Listing 9.1 realises the PCA for the 254 vectors of returns of the DAX components and generates Fig. 9.2.

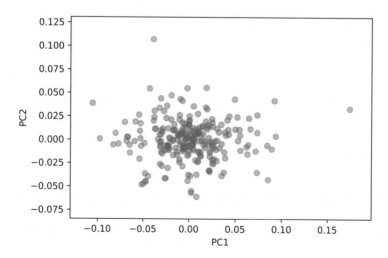

Fig. 9.2 PCA, for the 30 stocks of the DAX

Listing 9.1 Python. PCA for the 30 components

```
 1  # Library importations
 2  import pandas as pd
 3  import pandas_datareader.data as web
 4  import numpy as np
 5  from matplotlib import pyplot as plt
 6  fig, ax = plt.subplots()
 7
 8  # Data extraction
 9  data = pd.Series()
10  TickerDax = [ 'DPW.DE', 'ALV.DE', 'BMW.DE', 'DTE.DE', 'FME.DE', 'BAS.DE
        ',
11          'HEN3.DE', 'LIN.F', 'SAP.DE', 'DBK.DE', 'BAYN.DE', 'VOW3.DE',
12          'HEI.DE', 'FRE.F', 'MRK.DE', 'BEI.DE', 'SIE.DE', 'MUV2.DE',
13          'DB1.DE', 'VNA.DE', 'EOAN.F', 'DAI.DE', 'ADS.DE', 'WDI.F',
14          'RWE.DE', 'IFX.DE', '1COV.DE', 'TKA.DE', 'CON.DE', 'LHA.DE' ]
15  for x in TickerDax:
16  data[x] = web.DataReader(name = x, data_source='yahoo',start='
        2017-01-1',end='2017-12-31')
17  data_R = pd.DataFrame() # daily returns
18  for x in TickerDax:
19      data_R[x] = (data[x]['Close']/data[x]['Close'].shift(1)-1)
20
21  # PCA Analysis
22  from sklearn.decomposition import PCA
23  data_RC = pd.DataFrame()
24  data_RC = data_R.dropna() # the missing values are eliminated.
25  pca = PCA(n_components=2) # defining a 2-dimensional PCA
26  PC2 = pca.fit_transform(data_RC) # runs the PCA after centering
        the data: E(data_RC)=0
27
28  # Plot the projections of the daily observations of dimension 30
        on the two-dimensional PCA space
29  f_1 = PC2[:,0] # realisations factor 1
30  f_2 = PC2[:,1] # realisations factor 2
31  plt.scatter(f_1,f_2,alpha=.5)
32  plt.xlabel('PC1')
33  plt.ylabel('PC2')
```

In Fig. 9.2 we can visualise the 254 projections, or equivalently, the 254 realisations of the first two principal components (or factors) derived from the observations. From the program we get that the first principal component (factor) explains 28.84% of the variance (inertia), and the second principal component (factor) an additional 10.44%.

9.5.2 Explanation of the Factors for the DAX 30 Components

In the probabilistic model where b_1 and b_2 are orthonormal eigenvectors of maximum eigenvalues in the PCA decomposition of $\mathbf{Var}(R)$, let $RC = R - \mathbf{E}(R)$,

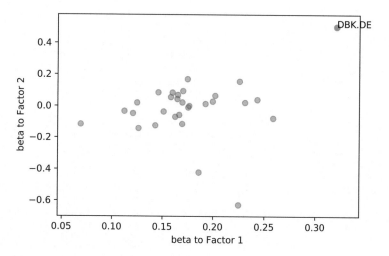

Fig. 9.3 Betas for the 30 stocks of the DAX 30

$f_1 = \langle RC, b_1 \rangle$, $f_2 = \langle RC, b_2 \rangle$ and $\epsilon = \sum_{i=3}^{30} \langle RC, b_i \rangle b_i$. Then the n vector of returns R can be decomposed as

$$R = \mathbf{E}(R) + b_1 f_1 + b_2 f_2 + \epsilon$$

and can be interpreted as a factor model.

If we call $\beta_1(i)$ the beta of the stock i to the first factor f_1 and call e_i the vector of $\mathbb{R}^{30}$ with all components equal to zero except for component i equal to 1 we get

$$\mathbf{Cov}(R_i, f_1) = \mathbf{Cov}(R_i, b_1'(R - \mathbf{E}(R)))$$

$$= e_i \mathbf{Var}(R) b_1 = \lambda_1 \langle e_i, b_1 \rangle \text{ and}$$

$$\mathbf{Var}(f_1) = b_1' \mathbf{Var}(R) b_1$$

$$= \lambda_1.$$

So, the beta of asset i to factor f_1 is $\beta_{f_1}(i) = \langle e_i, b_1 \rangle$.

Thus, the ith component of vector b_1 represents the beta of asset i to factor 1. As discussed in Proposition 9.3.2, these points should be inside a circle of radius 1. Then for all 30 stocks we represent their betas to the two factors f_1 and f_2 as in Fig. 9.3 by adding to Listing 9.1 the lines of codes appearing in Listing 9.2.

Listing 9.2 Python. Betas for the 30 stocks of the DAX

```
1  # This program is to be executed after Listing 9.1
2  # Plot the betas and the names of the stocks with maximum betas
3  proj1 = pca.components_.T[:,0] # components on the first axis
```

```
 4  proj2 = pca.components_.T[:,1] # components on the second axis
 5  indexproj1_max = np.argmax(proj1)
 6  indexproj2_max = np.argmax(proj2)
 7  Nameproj1_max = list(data_RC)[np.argmax(proj1)] # find the stock
        having the largest component on the first axis
 8  Nameproj2_max = list(data_RC)[np.argmax(proj2)] # find the stock
        having the largest component on the second axis
 9  fig, ax = plt.subplots()
10  plt.scatter(proj1, proj2, alpha=.5)
11  ax.scatter(proj1[indexproj1_max], proj2[indexproj1_max])
12  ax.annotate(Nameproj1_max,(proj1[indexproj1_max], proj2[
        indexproj1_max]))
13  ax.scatter(proj1[indexproj2_max], proj2[indexproj2_max])
14  ax.annotate(Nameproj2_max,(proj1[indexproj2_max], proj2[
        indexproj2_max]))
15  plt.xlabel('beta to Factor 1')
16  plt.ylabel('beta to Factor 2')
17  plt.show()
```

To explain the factors the "correlation circle" can also be plotted. In this representation we plot for each of the 30 stocks the points $(\rho(R_i, f_1), \rho(R_i, f_2))$ which represent the correlations of the returns of the stocks with the two factors. For this, the lines of codes of Listing 9.3 have to be added to Listing 9.2. We can see in Fig. 9.4 that all the points are within a circle of centre zero and radius 1, which is what was expected according to Property 9.3.3.

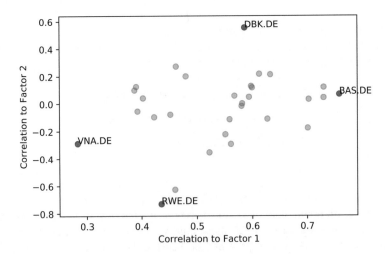

Fig. 9.4 Correlation circle for the 30 stocks of the DAX

Listing 9.3 Python. Correlation circle for the 30 stocks of the DAX

```
 1 # This program is to be executed after Listing 9.1 and Listing
      9.2
 2 # Correlation circle
 3 fig , ax = plt.subplots ()
 4 from scipy.stats.stats import pearsonr
 5 co = pd.DataFrame ()
 6 for sym in TickerDax :
 7 co[sym] = pearsonr(f_1 , data_RC[sym])[0] , pearsonr(f_2 , data_RC[
      sym ]) [0]
 8 plt.scatter(co.loc[0],co.loc[1],alpha =.5)
 9 # first point
10 indexcor1_max = np.argmax(co.loc[0].values)
11 indexcor1_max_x = co.loc[0][indexcor1_max]
12 indexcor1_max_y = co.loc[1][indexcor1_max]
13 indexcor1_max_name = list(co)[indexcor1_max]
14 ax.scatter(indexcor1_max_x,indexcor1_max_y,alpha=.5, color='red')
15 ax.annotate(indexcor1_max_name , (indexcor1_max_x,indexcor1_max_y
      ))
16 # second point
17 indexcor1_min = np.argmin(co.loc[0].values)
18 indexcor1_min_x = co.loc[0][indexcor1_min]
19 indexcor1_min_y = co.loc[1][indexcor1_min]
20 indexcor1_min_name = list(co)[indexcor1_min]
21 ax.scatter(indexcor1_min_x,indexcor1_min_y,alpha=.5, color='red')
22 ax.annotate(indexcor1_min_name , (indexcor1_min_x,indexcor1_min_y
      ))
23 # third point
24 indexcor2_max = np.argmax(co.loc[1].values)
25 indexcor2_max_x = co.loc[0][indexcor2_max]
26 indexcor2_max_y = co.loc[1][indexcor2_max]
27 indexcor2_max_name = list(co)[indexcor2_max]
28 ax.scatter(indexcor2_max_x,indexcor2_max_y,alpha=.5, color='red')
29 ax.annotate(indexcor2_max_name , (indexcor2_max_x,indexcor2_max_y
      ))
30 # fourth point
31 indexcor2_min = np.argmin(co.loc[1].values)
32 indexcor2_min_x = co.loc[0][indexcor2_min]
33 indexcor2_min_y = co.loc[1][indexcor2_min]
34 indexcor2_min_name = list(co)[indexcor2_min]
35 ax.scatter(indexcor2_min_x,indexcor2_min_y,alpha=.5, color='red')
36 ax.annotate(indexcor2_min_name , (indexcor2_min_x,indexcor2_min_y
      ))
37 plt.xlabel('Correlation to Factor 1')
38 plt.ylabel('Correlation to Factor 2')
```

A Few References

1. Ait-Sahalia, Y., & Xiu, D. (2017). Using principal component analysis to estimate a high dimensional factor model with high-frequency data. *Journal of Econometrics, 201*(2), 384–399.
2. Asness, C. S., Moskowitz, T. J., & Pedersen, L. J. (2013). Value and momentum everywhere. *The Journal of Finance, 68*(3), 929–985.
3. Bai, J.. & Ng, S. (2013, Sept). Principal components estimation and identification of static factors. *Journal of Econometrics, 176*(1), 18–29.
4. Connor, G., & Korajczyk, R. (1993). A test for the number of factors in an approximate factor model. *The Journal of Finance, 48*, 1263–1291.

Exercises and Problems

10

In this chapter some midterm and final exam subjects are presented with their solutions. Many of the subjects deal with alternative methods to demonstrate some important results from the course.

10.1 Midterm Exam, November 2015

Master M1: Mido 2nd November 2015 (Midterm Exam: Portfolio Management)

Exercise 1 We consider three risky financial assets S_1, S_2, S_3. We denote by S_0^i the value of S_i at time 0 and S_T^i its value at time T. R_i denotes the random return of S_i over $[0, T]$, i.e. $\frac{S_T^i}{S_0^i} = (1 + R_i)$, and we define

$$R = \begin{pmatrix} R_1 \\ R_2 \\ R_3 \end{pmatrix} \text{ and } M = \begin{pmatrix} m_1 \\ m_2 \\ m_3 \end{pmatrix} \text{ the vector of the expected returns } M := E[R].$$

Σ is the variance-covariance matrix of R, W_0 is the initial wealth, and $\alpha_i W_0$ is the money invested in asset S_i at time 0, for the period $[0, T]$. We allow short selling so α_i can be positive or negative. We call a strategy for which $\alpha_1 + \alpha_2 + \alpha_3 = 1$ an "investment strategy" and a strategy for which $\alpha_1 + \alpha_2 + \alpha_3 = 0$ a "self-financing strategy", and we denote by α' the transpose of α, so $\alpha' = (\alpha_1, \alpha_2, \alpha_3)$.

We denote by $W_T(\alpha)$ the value at time T of the portfolio built with the strategy α (so $W_0(\alpha) = W_0$ for an investment strategy and $W_0(\alpha) = 0$ for a self-financing strategy). We define, for both investment and self-financing strategies, $R(\alpha) = \frac{W_T(\alpha) - W_0(\alpha)}{W_0}$.

© Springer Nature Switzerland AG 2020
P. Brugière, *Quantitative Portfolio Management*, Springer Texts in Business and Economics, https://doi.org/10.1007/978-3-030-37740-3_10

(1.1)

   (a) Show that for any investment strategy or self-financing strategy α we have
$$R(\alpha) = \alpha' R.$$

   (b) Express $\mathbf{E}[R(\alpha)]$, $\mathbf{Var}[R(\alpha)]$, $\mathbf{E}[W_T(\alpha)]$ and $\mathbf{Var}[W_T(\alpha)]$ in terms of α, M, W_0 and Σ.

   From now on we assume that Σ is invertible and define $\langle x, y \rangle_{\Sigma^{-1}} := x'\Sigma^{-1}y$ and $\|x\|_{\Sigma^{-1}} := \sqrt{\langle x, x \rangle_{\Sigma^{-1}}}$. We also define the vector column $\mathbf{1} := (1, \cdots, 1)'$ and the scalars $b := \langle \mathbf{1}, M \rangle_{\Sigma^{-1}}$ and $a := \|\mathbf{1}\|_{\Sigma^{-1}}^2$.

(2.1)

   (a) Show that $\mathbf{Var}[R(\alpha)] = \|\Sigma\alpha\|_{\Sigma^{-1}}^2$.

   (b) Show that if $\alpha'\mathbf{1} = 1$ then $\langle \mathbf{1}, \Sigma\alpha \rangle_{\Sigma^{-1}} = 1$.

   (c) Show that for any investment strategy α we have $\mathbf{Var}[R(\alpha)] \geq \frac{1}{a}$ (you can use the Cauchy–Schwarz inequality).

(2.2) Calculate amongst all investment strategies α the strategy, denoted α_m, which minimises $\mathbf{Var}[R(\alpha)]$. Calculate $\mathbf{E}[R(\alpha_m)]$ and $\mathbf{Var}[R(\alpha_m)]$.

   We assume from now on that $M \neq \frac{b}{a}\mathbf{1}$.

(2.3)

   (a) Show that $\alpha_s = \dfrac{\Sigma^{-1}(M - \frac{b}{a}\mathbf{1})}{\|M - \frac{b}{a}\mathbf{1}\|_{\Sigma^{-1}}}$ is a self-financing strategy.

   (b) In this question, we assume that the random vector R is Gaussian. Show that the investment strategy $\alpha_e = \frac{\Sigma^{-1}\mathbf{1}}{a}$ and the self-financing strategy $\alpha_s = \dfrac{\Sigma^{-1}(M - \frac{b}{a}\mathbf{1})}{\|M - \frac{b}{a}\mathbf{1}\|_{\Sigma^{-1}}}$ have returns $R(\alpha_e)$ and $R(\alpha_s)$ which are independent random variables.

   (c) Calculate the variance of returns, $\mathbf{Var}[R(\alpha_s)]$, for the self-financing strategy
$$\alpha_s = \frac{\Sigma^{-1}(M - \frac{b}{a}\mathbf{1})}{\|M - \frac{b}{a}\mathbf{1}\|_{\Sigma^{-1}}}.$$

We assume from here that $M = (0.01, 0.03, 0.01)'$ and that
$$\Sigma = \begin{pmatrix} 0.1 & 0 & 0 \\ 0 & 0.1 & 0.1 \\ 0 & 0.1 & 0.2 \end{pmatrix}.$$

(3.1) Calculate the investment strategy α_m which minimises $\mathbf{Var}[R(\alpha)]$. Calculate $\mathbf{E}[R(\alpha_m)]$ and $\mathbf{Var}[R(\alpha_m)]$.

Exercise 2 We consider an investor with utility function $u(x) = 1 - \exp(-\lambda(x))$ with $\lambda > 0$. The investor has the choice between two investments I_1, I_2 of random returns R_1, R_2. The investor chooses the investment I_i that maximises $\mathbf{E}[u(R_i)]$.

We assume $R_1 \sim N(m_1, \sigma_1^2)$ and $R_2 \sim N(m_2, \sigma_2^2)$.

(1.1) Express $\mathbf{E}[u(R_i)]$, for $i = 1, 2$, in terms of m_i and σ_i^2.

(1.2) Determine C_i such that $u(C_i) = \mathbf{E}[u(R_i)]$ for $i = 1, 2$.

(1.3) What is the name of the quantity $\mathbf{E}[R_i] - C_i$ and what does it represent?

(1.4) Show that if $m_1 = m_2$ the investor will prefer I_1 to I_2 if and only if $\sigma_1 < \sigma_2$.

(1.5) Show that if $\sigma_1 = \sigma_2$ the investor will prefer I_1 to I_2 if and only if $m_1 > m_2$.

In this section v is a strictly increasing and concave function, and for any random variable X we define $\Pi_v(X) = \mathbf{E}[X] - v^{-1}(\mathbf{E}[v(X)])$. We recall that by Jensen's inequality, for v concave, $\mathbf{E}[v(X)] \leq v(\mathbf{E}[X])$.

(2.1) Show that for any X we have $\Pi_{vov}(X) \geq \Pi_v(X)$.

(2.2) When v is also twice differentiable calculate $-\frac{(vov)''}{(vov)'}$ and $-\frac{v''}{v'}$. Which one is the biggest? Why was this predictable from 2.1)?

(2.3) Calculate $-\frac{(uou)''}{(uou)'}$ and $-\frac{u''}{u'}$ for $u(x) = 1 - \exp(-\lambda(x))$.

10.1.1 Solutions: Midterm Exam, November 2015

Master M1: Mido 2015–2016 (Midterm Exam: Portfolio Management)

Exercise 1

(1.1)

(a) $\alpha_i W_0$ invested in asset $i \implies \frac{\alpha_i W_0}{S_0^i}$ asset i held.

For both investing and self-financing portfolios $W_T(\alpha) = \sum_{i=1}^{3} \alpha_i \frac{W_0}{S_0^i} S_T^i$ and

$$W_0(\alpha) = \sum_{i=1}^{3} \alpha_i W_0 \implies \frac{W_T(\alpha) - W_0(\alpha)}{W_0} = \sum_{i=1}^{3} \alpha_i (\frac{S_T^i}{S_0^i} - 1) = \sum_{i=1}^{3} \alpha_i R_i = \alpha' R.$$

(b)

- $\mathbf{E}[\alpha' R] = \alpha' \mathbf{E}[R] = \alpha' M.$
- $\mathbf{Var}[\alpha' R] = \alpha' \mathbf{Var}[R]\alpha = \alpha' \Sigma \alpha.$
- $W_T(\alpha) = W_0 \alpha' R + W_0(\alpha) \implies \mathbf{E}(W_T(\alpha)) = W_0(\alpha) + W_0 \alpha' M.$
  So, $\mathbf{E}(W_T(\alpha)) = W_0 \alpha' M$ if α is a self-financing strategy and $\mathbf{E}(W_T(\alpha)) = W_0(1 + \alpha' M)$ if α is an investment strategy.
- $\mathbf{Var}[W_0 \alpha' R + W_0(\alpha)] = W_0^2 \mathbf{Var}[\alpha' R] = W_0^2 \alpha' \Sigma \alpha.$

(2.1)

(a) $\Sigma' = \Sigma$ and $\mathbf{Var}[R(\alpha)] = \alpha'\Sigma\alpha = (\Sigma\alpha)'(\Sigma^{-1}\Sigma)\alpha = \|\Sigma\alpha\|^2_{\Sigma^{-1}}$.

(b) $\langle \mathbf{1}, \Sigma\alpha\rangle_{\Sigma^{-1}} = \mathbf{1}'\Sigma^{-1}\Sigma\alpha = \mathbf{1}'\alpha = 1$.

(c) $\langle \cdot, \cdot\rangle_{\Sigma^{-1}}$ is a scalar product. Cauchy–Schwarz implies $\langle \mathbf{1}, \Sigma\alpha\rangle_{\Sigma^{-1}} \leq$ $\|\mathbf{1}\|_{\Sigma^{-1}}\|\Sigma\alpha\|_{\Sigma^{-1}}$ so $\|\Sigma\alpha\|_{\Sigma^{-1}} \geq \frac{\langle \mathbf{1}, \Sigma\alpha\rangle_{\Sigma^{-1}}}{\|\mathbf{1}\|_{\Sigma^{-1}}} = \frac{1}{\sqrt{a}}$, therefore $\mathbf{Var}[R(\alpha)] \geq \frac{1}{a}$.

(2.2)

- Minimum of $\alpha'\Sigma\alpha$ under the constraint $\alpha'\mathbf{1} = 1$.
  $\frac{\partial L}{\partial \alpha} = 0$ implies $\alpha = \frac{\lambda}{2}\Sigma^{-1}\mathbf{1}$ and the constraint implies $\frac{\lambda}{2} = \frac{1}{\mathbf{1}'\Sigma^{-1}\mathbf{1}} = \frac{1}{a}$.
- The function is strictly convex, so this is an extremum and $\alpha_m = \frac{1}{a}\Sigma^{-1}\mathbf{1}$.
- $\mathbf{E}[R(\alpha_m)] = \alpha'_m M = \frac{1}{a}\langle \mathbf{1}, M\rangle_{\Sigma^{-1}} = \frac{b}{a}$.
- $\mathbf{Var}[R(\alpha_m)] = \alpha'_m \Sigma\alpha_m = \frac{1}{a}$.

(2.3)

(a) $\mathbf{1}'\Sigma^{-1}(M - \frac{b}{a}\mathbf{1}) = b - \frac{b}{a}a = 0$ so self-financing

(b) $(R(\alpha_e), R(\alpha_s))$ is an affine transformation of $R(\alpha)$ and therefore is a Gaussian vector. So, to prove independence it is enough to show that the covariance is zero.
$(\Sigma^{-1}\mathbf{1})'\Sigma\Sigma^{-1}(M - \frac{b}{a}\mathbf{1}) = b - \frac{b}{a}a = 0$ which proves the result.

(c) $(\Sigma^{-1}(M - \frac{b}{a}\mathbf{1}))'\Sigma\Sigma^{-1}(M - \frac{b}{a}\mathbf{1}) = \|M - \frac{b}{a}\mathbf{1}\|^2_{\Sigma^{-1}}$ so the variance is 1.

(3.1) The matrix can be inverted by blocks

$$\Sigma^{-1} = \begin{pmatrix} 10 & 0 & 0 \\ 0 & 20 & -10 \\ 0 & -10 & 10 \end{pmatrix}$$

$a = 20$, $\alpha_m = (0.5, 0.5.0)'$, $b = 0.4$, $\mathbf{E}[R(\alpha_m)] = \frac{b}{a} = 0.02$, $\mathbf{Var}[R(\alpha_m)] = \frac{1}{a} = 0.05$.

Exercise 2

(1.1) $\mathbf{E}[u(R_i)] = 1 - \exp(-\lambda(m_i + \sigma_i Z))$ with $Z \sim N(0, 1)$,
so $\mathbf{E}[u(R_i)] = 1 - \exp(-\lambda m_i)\exp(-\lambda\sigma_i Z) = 1 - \exp(-\lambda m_i)\exp(\frac{(\lambda\sigma_i)^2}{2})$.

(1.2) $u(C_i) = \mathbf{E}[u(R_i)] \Leftrightarrow 1 - \exp(-\lambda C_i) = 1 - \exp(-\lambda m_i)\exp(\frac{(\lambda\sigma_i)^2}{2})$
$\Leftrightarrow C_i = m_i - \frac{\lambda\sigma_i^2}{2}$.

(1.3) Risk premium: discount to the "risk neutral value" that must be given to the risky investor.

(1.4) and (1.5) Conditions for $m_1 - \frac{\lambda}{2}\sigma_1^2 > m_2 - \frac{\lambda}{2}\sigma_2^2$ (obvious).

(2.1) $\Pi_{vov} \geq \Pi_v \Leftrightarrow \mathbf{E}[X] - (v \circ v)^{-1}(\mathbf{E}[(v \circ v)(X)]) \geq \mathbf{E}[X] - v^{-1}(\mathbf{E}[v(X)])$

$\Leftrightarrow (v \circ v)^{-1}(\mathbf{E}[(v \circ v)(X)]) \leq v^{-1}(\mathbf{E}[v(X)])$

$\Leftrightarrow (\mathbf{E}[(v \circ v)(X)]) \leq (v \circ v)(v^{-1}(\mathbf{E}[v(X)]))$ (because $v \circ v$ is increasing)

$\Leftrightarrow (\mathbf{E}[(v \circ v)(X)]) \leq v(\mathbf{E}[v(X)])$ which is true because v is concave.

(2.2) $(v \circ v)' = (v' \circ v)v'$ and $(v \circ v)'' = (v'' \circ v)v'^2 + (v' \circ v)v'' - \frac{(vov)''}{(vov)'} =$

$-\frac{(v''ov)v'^2}{(v'ov)v'} - \frac{(v'ov)v''}{(v'ov)v'} = -\frac{(v''ov)v'}{(v'ov)} - \frac{v''}{v'}$, so $-\frac{(vov)''}{(vov)'} \geq -\frac{v''}{v'}$ as $-v'' \geq 0$

(because v concave) and $v' > 0$ (because v is strictly increasing). (2.1) shows that vov is more risk adverse, concave, than vov, so the risk aversion indicator for $v \circ v$ is higher than the one for v.

(2.3) $u'(x) = \lambda \exp(-\lambda x)$ and $u''(x) = -\lambda^2 \exp(-\lambda x)$ so $-\frac{(vov)''}{(vov)'} =$

$\lambda^2 \exp(-\lambda x) + \lambda$ and $-\frac{v''}{v'} = \lambda$.

10.2   Exam, January 2016

Master M1: Mido 5th January 2016 (Exam: Portfolio Management[1]: Time 1 h 30 min)

Exercise 1 We consider d risky assets S_1, S_2, ...S_d, whose returns satisfy $r_i = m_i + \epsilon_i$. Matricially:

$$R = M + \epsilon \text{ with } R = \begin{pmatrix} r_1 \\ \vdots \\ r_d \end{pmatrix}, M = \begin{pmatrix} m_1 \\ \vdots \\ m_d \end{pmatrix} \text{ and } \epsilon = \begin{pmatrix} \epsilon_1 \\ \vdots \\ \epsilon_d \end{pmatrix},$$

where M is a constant and ϵ is a random vector with zero expectation and variance-covariance matrix $\sigma^2 \mathrm{Id}_{R^d}$. We also assume that there is a risk-free asset of expected return m_0 and we denote by π_M the risky investment portfolio which defines with the risk-free asset the "Capital Market Line" (which is tangent to the efficient frontier $\mathcal{F}$). We define $\pi'_M = (\pi_M^1, \cdots, \pi_M^d)$.

(1) Express r_M as a function of π_M, M and ϵ.
(2) Calculate $\mathbf{Var}(r_M)$ as a function of π_M and σ.
(3) Calculate $\mathbf{Cov}(r_M, r_i)$ as a function of π_M, π_M^i and σ.
   We recall the equation of the Security Market Line ("SML"):

$$m_P - m_0 = \beta_M(P)(m_M - m_0)$$

(where $\beta_M(P) = \mathbf{Cov}(r_P, r_M)/\mathbf{Var}(r_M)$).

---

[1]Pierre Brugière, University Paris 9 Dauphine.

(4) Calculate $\beta_M(i)$ for all assets i as a function of π_M^i.

(5)

    (a) From the SML equation deduce the expression of π_M as a function of the m_i and m_0 when $\frac{1}{d}\sum_{i=1}^d m_i \neq m_0$.

    (b) What happens when $\frac{1}{d}\sum_{i=1}^d m_i = m_0$?

Exercise 2 We consider d risky assets $S_1, S_2, \ldots S_d$. Let π_F be an investment portfolio built from the S_i, which belongs to the efficient frontier $\mathcal{F}$ defined for the risky assets, and let π_P and π_Q be two investment portfolios built from the S_i. For $\lambda \in \mathbb{R}$ we define the portfolios $\pi_{P,\lambda} = \lambda \pi_F + (1-\lambda)\pi_P$ and $\pi_{Q,\lambda} = \lambda \pi_F + (1-\lambda)\pi_Q$ and we use the following notations:

$$m_P = \mathbf{E}(R(\pi_P)), m_Q = \mathbf{E}(R(\pi_Q)),\ m_{P,\lambda} = \mathbf{E}(R(\pi_{P,\lambda})),\ m_{Q,\lambda} = \mathbf{E}(R(\pi_{Q,\lambda})),$$

$$\sigma_P = \sigma(R(\pi_P)),\ \sigma_Q = \sigma(R(\pi_Q)),\ \sigma_{P,\lambda} = \sigma(R(\pi_{P,\lambda})),\ \sigma_{Q,\lambda} = \sigma(R(\pi_{Q,\lambda})),$$

$$m_F = \mathbf{E}(R(\pi_F)),\ \sigma_F = \sigma(R(\pi_F)).$$

(1) Express $m_{P,\lambda}$ and $m_{Q,\lambda}$ as a function of λ, m_F, m_P and m_Q.

(2) Express $\sigma_{P,\lambda}$ and $\sigma_{Q,\lambda}$ as a function of λ, σ_F, σ_P and σ_Q and of the correlations for the returns $\rho_{P,F}$ and $\rho_{Q,F}$.

(3) What type of curve is the curve $\left\{ \begin{pmatrix} \sigma_{Q,\lambda} \\ m_{Q,\lambda} \end{pmatrix}, \lambda \in [0,1] \right\}$ in the following cases:

    (a) $\rho_{Q,F} = 1$,

    (b) $\rho_{Q,F} = -1$,

    (c) $-1 < \rho_{Q,F} < 1$?

(4) Calculate the vectors $\left(\frac{d}{d\lambda}\right)_{\lambda=1}\begin{pmatrix} \sigma_{P,\lambda} \\ m_{P,\lambda} \end{pmatrix}$ and $\left(\frac{d}{d\lambda}\right)_{\lambda=1}\begin{pmatrix} \sigma_{Q,\lambda} \\ m_{Q,\lambda} \end{pmatrix}$.

(5) Why should the two vectors defined in (4) be collinear?

(6) Deduce from (5) a relationship between m_F, m_P, m_Q, $\beta_F(P)$ and $\beta_F(Q)$.

    We assume now that there is a risk-free asset of return m_0 and denote by π_F the investment portfolio made of the risky assets which defines with the risk-free portfolio the "Capital Market Line". Let Π_λ be the investment portfolio built by investing the proportion λ of the wealth in π_F and $1 - \lambda$ in the risk-free asset. We define $m_\lambda = \mathbf{E}(R(\Pi_\lambda))$ and $\sigma_\lambda = \sigma(R(\Pi_\lambda))$.

(7) Give the expressions of m_λ and σ_λ as functions of λ, m_F, m_0 and σ_F.

(8) Calculate $\left(\frac{d}{d\lambda}\right)_{\lambda=1}\begin{pmatrix} \sigma_\lambda \\ m_\lambda \end{pmatrix}$.

(9) Deduce from the previous results the "Security Market Line" equation

$$m_P - m_0 = \beta_F(P)(m_F - m_0).$$

Exercise 3 We consider a two-factor model with factors F^1, F^2:

$$R^i = \alpha^i + \beta_1^i F^1 + \beta_2^i F^2 + \epsilon^i, \quad i = 1, \cdots, d.$$

We assume that $\mathbf{E}(F^1) = 0, \mathbf{E}(F^2) = 0$ and that $\mathbf{Cov}(\epsilon^i, F^j) = 0$ for all $i = 1, \cdots, d$ and $j = 1, 2$. We denote by Σ_F the variance-covariance matrix for the F^i, i.e. $\Sigma_F := (\mathbf{Cov}(F^i, F^j))$. We assume that the APT conditions are satisfied, so:

$$\exists \lambda^0, \lambda^1, \lambda^2 \text{ such that: } \forall i \in [\![1, d]\!], \mathbf{E}(R^i) = \lambda^0 + \beta_1^i \lambda^1 + \beta_2^i \lambda^2.$$

We assume that:

Asset i	$\mathbf{E}(R^i)$	β_1^i	β_2^i
1	8%	2	1
2	9%	1	2
3	8%	−1	3

(1) Calculate $\lambda^0, \lambda^1, \lambda^2$.

   Let π be an investment portfolio whose sensitivities to the factors F^1 and F^2 are $\beta_1^\pi = \frac{1}{2}$ and $\beta_2^\pi = 2$.

(2)

   (a) Calculate the return of this portfolio.
   (b) Determine the standard deviation of the returns of the portfolio π knowing that its specific risk is 20% and that

$$\Sigma_F = \begin{pmatrix} 0.04 & 0.02 \\ 0.02 & 0.04 \end{pmatrix}.$$

   We recall that a symmetric matrix can be diagonalised, i.e. that there exists an orthogonal P (i.e. such that $P'P = PP' = \mathrm{Id}$) such that $P^{-1}SP$ is diagonal.

(3)

   (a) Why is it possible to rewrite the factor model defined by F^1 and F^2 as a factor model where the factors G^1 and G^2 have a diagonal variance-covariance matrix Σ_G?
   (b) Express the factors G^1 and G^2 of 3(a) as a function of F^1 and F^2. If we consider asset 1 which satisfies $R^1 = \alpha^1 + 2F^1 + F^2 + \epsilon^1$, find γ_1^1 and γ_2^1 such that $R^1 = \alpha^1 + \gamma_1^1 G^1 + \gamma_2^1 G^2 + \epsilon^1$.

10.2.1 Solutions: Exam, January 2016

Master M1: Mido 5th January 2016 (Exam: Portfolio Management[2])

Exercise 1

(1) $r_M = \pi'_M M + \pi'_M \epsilon$.

(2) $\mathbf{Var}(r_M) = \mathbf{Var}(\pi'_M M + \pi'_M \epsilon) = \mathbf{Var}(\pi'_M \epsilon) = \sigma^2 \|\pi_M\|^2$.

(3) $\mathbf{Cov}(r_M, r_i) = \sigma^2 \pi^i_M$.

(4) $\beta_M(i) = \dfrac{\sigma^2 \pi^i_M}{\sigma^2 \|\pi_M\|^2} = \dfrac{\pi^i_M}{\|\pi_M\|^2}$.

(5)

   (a) $m_i - m_0 = \beta_M(i)(m_M - m_0)$

$$\Rightarrow m_i - m_0 = \frac{\pi^i_M}{\|\pi_M\|^2}(m_M - m_0)$$

$$\Rightarrow \sum_{i=1}^{i=d}(m_i - m_0) = \frac{1}{\|\pi_M\|^2}(m_M - m_0)$$

$$\Rightarrow m_i - m_0 = \pi^i_M \sum_{i=1}^{i=d}(m_i - m_0),$$

so, if $\sum_{i=1}^{i=d}(m_i - m_0) \neq 0$ then $\pi^i_M = \dfrac{m_i - m_0}{\sum_{i=1}^{i=d}(m_i - m_0)}$.

   (b) It is easy to see that the minimum variance portfolio satisfies $\pi^i = \frac{1}{d}$ so, if $\frac{1}{d}\sum_{i=1}^{d} m_i = m_0$ then $\frac{b}{a} = m_0$ and in this case there is no tangent portfolio π_M!

Exercise 2

(1) $m_{P,\lambda} = \lambda m_F + (1 - \lambda)m_P, \, m_{Q,\lambda} = \lambda m_F + (1 - \lambda)m_Q$.

(2) $\sigma_{P,\lambda} = \left(\lambda^2 \sigma_F^2 + (1 - \lambda)^2 \sigma_P^2 + 2\rho_{P,F}\lambda(1 - \lambda)\sigma_F \sigma_P\right)^{\frac{1}{2}}$ and

$\sigma_{Q,\lambda} = \left(\lambda^2 \sigma_F^2 + (1 - \lambda)^2 \sigma_Q^2 + 2\rho_{Q,F}\lambda(1 - \lambda)\sigma_F \sigma_Q\right)^{\frac{1}{2}}$.

(3)

   (a) It is a segment between the points $(\sigma_{F,\lambda}, m_{F,\lambda})$ and $(\sigma_{Q,\lambda}, m_{Q,\lambda})$.

   (b) It is a cone joining the points $(\sigma_{F,\lambda}, m_{F,\lambda})$ and $(\sigma_{Q,\lambda}, m_{Q,\lambda})$ and which intersects the axis $\sigma = 0$.

   (c) It is a portion of a hyperbola which passes through the points $(\sigma_{F,\lambda}, m_{F,\lambda})$ and $(\sigma_{Q,\lambda}, m_{Q,\lambda})$.

(4) $\left(\dfrac{d}{d\lambda}\right)_{\lambda=1} \left(\lambda^2 \sigma_F^2 + (1 - \lambda)^2 \sigma_P^2 + 2\rho_{P,F}\lambda(1 - \lambda)\sigma_F \sigma_P\right)^{\frac{1}{2}} = \dfrac{2\sigma_F^2 - 2\rho_{P,F}\sigma_F \sigma_P}{2\sigma_F}$, so

$\left(\dfrac{d}{d\lambda}\right)_{\lambda=1} \left(\dfrac{\sigma_{P,\lambda}}{m_{P,\lambda}}\right) = \left(\dfrac{\sigma_F - \rho_{P,F}\sigma_P}{m_F - m_P}\right)$ and we get the same expressions for Q.

---

[2]Pierre Brugière, University Paris 9 Dauphine.

(5) The two vectors must be tangent to the efficient frontier $\mathcal{F}$ and therefore are collinear.

(6) We must have $(\sigma_F - \rho_{P,F}\sigma_P)(m_F - m_Q) = (\sigma_F - \rho_{Q,F}\sigma_Q)(m_F - m_P)$ and therefore $(1 - \beta_{P,F})(m_F - m_Q) = (1 - \beta_{Q,F})(m_F - m_P)$.

(7) $\sigma_\lambda = |\lambda|\sigma_F$ and $m_\lambda = \lambda m_F + (1 - \lambda)m_0$.

(8) $\left(\frac{d}{d\lambda}\right)_{\lambda=1} \begin{pmatrix} \sigma_\lambda \\ m_\lambda \end{pmatrix} = \begin{pmatrix} \sigma_F \\ m_F - m_0 \end{pmatrix}.$

(9) The derivative vectors calculated in (4) and (8) define both tangent vectors to the efficient frontier and so must be collinear, so the determinant between $\begin{pmatrix} \sigma_F \\ m_F - m_0 \end{pmatrix}$ and $\begin{pmatrix} \sigma_F - \rho_{P,F}\sigma_P \\ m_F - m_P \end{pmatrix}$ must be zero, so

$$\sigma_F(m_F - m_P) = (\sigma_F - \rho_{P,F}\sigma_P)(m_F - m_0)$$

$$\Rightarrow \sigma_F(m_0 - m_P) = -\rho_{P,F}\sigma_P(m_F - m_0)$$

$$\Rightarrow (m_P - m_0) = \rho_{P,F}\frac{\sigma_P}{\sigma_F}(m_F - m_0).$$

Exercise 3

(1) Three equations for three unknowns solved with $m_0 = 1\%$, $\lambda_1 = 2\%$, $\lambda_3 = 3\%$.

(2)

   (a) Expected return: $1\% + 0.5 \times 2\% + 2 \times 3\% = 8\%$.

   (b) Return variance: variance linked to the factors + specific variance $\frac{1}{4} \times 4\% + 4 \times 4\% + 2 \times 2\% + 4\% = 25\%$. So, the standard deviation is 50%.

(3)

   (a) The model is $R = A + BF + \epsilon$. Let P be orthogonal ($P'P = \text{Id}$) and such that $P'\Sigma_F P = D$ then $R = A + BF + \epsilon \Rightarrow R = A + BPP'F + \epsilon = A + (BP)(P'F) + \epsilon$. Let $G = P'F$ then $\mathbf{Var}(G) = P'\mathbf{Var}(F)P = D$.

   (b) The eigenvalues are 2% and 6% and the eigenvectors are $(\frac{1}{\sqrt{2}}, \frac{-1}{\sqrt{2}})'$ for the eigenvalue 2% and $(\frac{1}{\sqrt{2}}, \frac{1}{\sqrt{2}})'$ for the eigenvalue 6%.
   $G = P'F \Rightarrow F = PG$ so $R^1 = \alpha^1 + 2 \times (\frac{1}{\sqrt{2}}G^1 + \frac{1}{\sqrt{2}}G^2) + (-\frac{1}{\sqrt{2}}G^1 + \frac{1}{\sqrt{2}}G^2)$
   so, $\gamma_{1,1} = \frac{1}{\sqrt{2}}$ and $\gamma_{1,2} = \frac{3}{\sqrt{2}}$.

10.3   Midterm Exam, November 2016

Master M1: Mido 3rd November 2016 (Exam: Portfolio Management[3]: Time 2 h)

Notations We consider d risky assets $S_1, S_2, \cdots, S_d$ whose returns over $[0, T]$ satisfy $R^i = m^i + \epsilon^i$. We write matricially

$$R = M + \epsilon \text{ with } R = \begin{pmatrix} R^1 \\ \vdots \\ R^d \end{pmatrix}, \ M = \begin{pmatrix} m^1 \\ \vdots \\ m^d \end{pmatrix} \text{ and } \epsilon = \begin{pmatrix} \epsilon^1 \\ \vdots \\ \epsilon^d \end{pmatrix},$$

where M is a vector in $\mathbb{R}^d$ and ϵ is a random vector of expectation zero and variance-covariance matrix Σ which is invertible. We denote by: $\pi' = (\pi^1, \cdots, \pi^d)$ the proportion at time zero of the investments in the risky assets, S_i, R^π the return of the portfolio over $[0, T]$, m^π its expected return and σ^π its standard deviation. The vector in $\mathbb{R}^d$ whose components are all 1 is denoted 1_d. For a function $L(\pi)$ we denote by $\frac{\partial L}{\partial \pi}$ the line vector $(\frac{\partial L}{\partial \pi^1}, \cdots, \frac{\partial L}{\partial \pi^d})$.

Problem   **[13pts]**
For all $\lambda > 0$ we want to solve:

$$(P_\lambda) \begin{cases} \sup\limits_{\pi \in \mathbb{R}^d} \ \mathbf{E}(R^\pi) - \lambda \mathbf{Var}(R^\pi) \\ \qquad \pi' 1_d = 1 \end{cases}$$

(1)
    (a) Why is this problem interesting?   **[0.50pt]**
    (b) For an investment portfolio π, give without proof the expression of R^π as a function of π and R.   **[0.50pt]**
    (c) Express $\mathbf{E}(R^\pi)$ as a function of π and M.   **[0.50pt]**
    (d) Express $\mathbf{Var}(R^\pi)$ as a function of π and Σ.   **[0.50pt]**
    (e) Without any calculation explain why the sup of (P_λ) is reached.   **[0.50pt]**
        We denote by π_λ the allocation solution π of (P_λ).

(2)
    (a) Recall the mathematical definition of an efficient portfolio.   **[0.50pt]**
    (b) Prove that π_λ is an efficient portfolio.   **[0.50pt]**
        For all x, y in $\mathbb{R}^d$ we denote by $\langle x, y \rangle_{\Sigma^{-1}}$ the quantity $x' \Sigma^{-1} y$ and by $\|x\|_{\Sigma^{-1}}$ the quantity $\sqrt{\langle x, x \rangle_{\Sigma^{-1}}}$.

---

[3]Pierre Brugière, University Paris 9 Dauphine.

(3)

- (a) Prove that Σ is symmetric. **[0.50pt]**
- (b) Prove that Σ is positive. **[0.50pt]**
- (c) Without using the fact that $\Sigma = \Sigma'$, show that $\forall x \in \mathbb{R}^d, x'\Sigma^{-1}x = x'(\Sigma^{-1})'x$. **[0.50pt]**
- (d) Prove that $(\Sigma^{-1})' = \Sigma^{-1}$. **[0.50pt]**
- (e) By studying the minimum of $f : x \longrightarrow x'\Sigma x$ show that Σ invertible $\Rightarrow \forall x \in \mathbb{R}^d \setminus \{0\}\ x'\Sigma x > 0$. **[0.50pt]**
- (f) Without any calculation explain why $\langle \cdot, \cdot \rangle_{\Sigma^{-1}}$ is a scalar product on $\mathbb{R}^d$. **[0.50pt]**

From now on we also assume that M is not collinear to 1_d and we define:
$a = \langle 1_d, 1_d \rangle_{\Sigma^{-1}}, b = \langle M, 1_d \rangle_{\Sigma^{-1}}$ and $c = \langle M - \frac{b}{a}1_d, M - \frac{b}{a}1_d \rangle_{\Sigma^{-1}}$.

(4)

- (a) Prove that $\langle M - \frac{b}{a}1_d, 1_d \rangle_{\Sigma^{-1}} = 0$. **[0.5pt]**
- (b) Express $\|M\|^2_{\Sigma^{-1}}$ as a function of a, b and c. **[0.50pt]**

We recall that the Lagrangian for (P_λ) is:

$$L_\lambda(\pi, \mu) = \mathbf{E}(R^\pi) - \lambda \mathbf{Var}(R^\pi) - \mu(\pi'1_d - 1).$$

(5)

- (a) Show that: $\frac{\partial L_\lambda}{\partial \pi} = 0 \Leftrightarrow \pi = \frac{1}{2\lambda}\Sigma^{-1}(M - \mu 1_d)$. **[0.50pt]**
- (b) Calculate the expression of π_λ. **[1.00pt]**
- (c) Express $\mathbf{E}(R^{\pi_\lambda})$ as a function of a, b, c and λ. **[1.00pt]**
- (d) Express $\mathbf{Var}(R^{\pi_\lambda})$ as a function of a, b, c and λ. **[1.00pt]**
- (e) What is the value of $\lim_{\lambda \to +\infty} \pi_\lambda$? What can you say about this portfolio? **[0.50pt]**
- (f) What can you say about $\mathcal{F}^+ = \{\pi_\lambda, \lambda > 0\}$? **[0.50pt]**

For all $\gamma > 0$, we search for the portfolio solution of

$$(P^*_\gamma) \begin{cases} \inf_{\pi \in \mathbb{R}^d} \mathbf{Var}(R^\pi) - \gamma \mathbf{E}(R^\pi) \\ \pi'1_d = 1. \end{cases}$$

Such a solution is denoted π^*_γ.

(6)

- (a) Prove that $\forall \gamma > 0, \pi^*_\gamma = \pi_{\frac{1}{\gamma}}$. **[1.00pt]**
- (b) What is the difference of approach between solving (P^*_γ) and (P_λ)? **[1.00pt]**

Exercise (Sharpe Ratio) **[7pts]**

We assume here that there is a risk-free asset S_0 of return r_0 over $[0, T]$ which is available. We assume that $M \neq r_0 1_d$. An investment portfolio is defined by its vector of allocation π in the risky assets. The investment in the risk-free asset for such a portfolio is $1 - \pi'1_d$. Here we search for investment portfolio solutions of
(S): $\sup_{\pi \in \mathbb{R}^d} \frac{\mathbf{E}(R^\pi - r_0)}{\sigma^\pi}$.

(1)

    (a) Show that $R^\pi = \pi'(R - r_0 1_d) + r_0$.   **[0.50pt]**

    (b) Express $\mathbf{E}(R^\pi - r_0)$ as a function of π, M, r_0 and Σ.   **[0.50pt]**

    (c) Express σ^π as a function of π and Σ.   **[0.50pt]**

    We consider the problem (V): $\displaystyle\sup_{\pi \in \mathbb{R}^d} \frac{\left(\mathbf{E}(R^\pi - r_0)\right)^2}{\mathbf{Var}(R^\pi)}$.

(2)

    (a) Show that π reaches the max for (S) $\Leftrightarrow$ π reaches the max for (V). **[0.5pt]**

    (b) Express $\dfrac{\partial\left(\mathbf{E}(R^\pi - r_0)\right)^2}{\partial \pi}$ as a function of π, M and r_0.   **[1.00pt]**

    (c) Express $\dfrac{\partial \mathbf{Var}(R^\pi)}{\partial \pi}$ as a function of π and Σ.   **[0.50pt]**

    (d) Show that if $E(R^\pi - r_0) \neq 0$ then   **[1.00pt]**

$$\frac{\partial}{\partial \pi} \frac{\left(\mathbf{E}(R^\pi - r_0)\right)^2}{\mathbf{Var}(R^\pi)} = 0 \Leftrightarrow \mathbf{Var}(R^\pi)(M - r_0 1_d) - \left(\mathbf{E}(R^\pi - r_0)\right)\Sigma\pi = 0.$$

If necessary, you can admit the results of (2) to show (3).

(3)

    (a) Show that there exists an investment portfolio R^π such that $\frac{E(R^\pi - r_0)}{\sigma^\pi} > 0$.   **[0.50pt]**

    (b) Show that the solutions of (S) are the portfolios $\alpha \Sigma^{-1}(M - r_0 1_d)$ with $\alpha \in \mathbb{R} \setminus \{0\}$. For these portfolios, what is the value of $\frac{E(R^\pi - r_0)}{\sigma^\pi}$?   **[1.00pt]**

    (c) Show that if $\frac{b}{a} \neq r_0$ there exists a unique investment portfolio made exclusively of risky assets which is a solution of (S). Calculate π for this portfolio.   **[1.00pt]**

10.3.1 Solutions: Midterm Exam, November 2016

Problem

(1)

    (a) Because it makes it possible to maximise the expected return while giving a cost to the risk. It can be seen as a Lagrangian for a maximisation problem of the expectation under a risk constraint.

    (b) $R^\pi = \pi' R$.

    (c) $R^\pi = \pi' M$.

    (d) $\mathbf{Var}(R^\pi) = \pi' \Sigma \pi$.

    (e) Σ is invertible and $\lambda > 0$ so the negative quadratic term $-\lambda \mathbf{Var}(R^\pi)$ dominates and its limit is $-\infty$ as $\|\pi\|$ tends to $+\infty$. So, the Sup is to be found within a ball of $\mathbb{R}^d$ and a continuous function on a compact set reaches its maximum.

(2)

(a) π^* is efficient iff $\forall \pi, \mathbf{E}(R^\pi) \geq \mathbf{E}(R^{\pi^*}) \Rightarrow \mathbf{Var}(R^\pi) \geq \mathbf{Var}(R^{\pi^*})$.

(b) $\mathbf{E}(R^{\pi^*}) - \lambda \mathbf{Var}(R^{\pi^*}) \geq \mathbf{E}(R^\pi) - \lambda \mathbf{Var}(R^\pi)$
$\Rightarrow \mathbf{Var}(R^\pi) - \mathbf{Var}(R^{\pi^*}) \geq \frac{1}{\lambda}(\mathbf{E}(R^\pi) - \mathbf{E}(R^{\pi^*}))$, which proves the result.

(3)

(a) $\mathbf{Cov}(R^i, R^j) = \mathbf{E}([R^i - \mathbf{E}(R^i)][R^j - \mathbf{E}(R^j)]) = \mathbf{E}([R^j - \mathbf{E}(R^j)][R^i - \mathbf{E}(R^i)])$
$= \mathbf{Cov}(R^j, R^i)$.

(b) $\forall x \in \mathbb{R}^d, x'\Sigma x = \mathbf{Var}(x'R) = \mathbf{E}([x'R - \mathbf{E}(x'R)]^2) \geq 0$.

(c) $x'\Sigma^{-1}x$ is a number so is equal to its transpose and $(x')' = x$.

(d) $\Sigma\Sigma^{-1} = \Sigma^{-1}\Sigma = \mathrm{Id}_{\mathbb{R}^d} \Rightarrow (\Sigma\Sigma^{-1})' = (\Sigma^{-1}\Sigma)' = \mathrm{Id}_{\mathbb{R}^d}$
$\Rightarrow (\Sigma^{-1})'\Sigma' = \Sigma'(\Sigma^{-1})' = \mathrm{Id}_{\mathbb{R}^d} \Rightarrow (\Sigma^{-1})'\Sigma = \Sigma(\Sigma^{-1})' = \mathrm{Id}_{\mathbb{R}^d}$
$\Rightarrow (\Sigma^{-1})' = \Sigma^{-1}$.

(e) If $x'\Sigma x = 0$ then a minimum is reached for f in x, so $\frac{\partial f}{\partial x} = 0$, hence $\Sigma x = 0$ and as Σ is invertible, $x = 0$. Therefore, $x \neq 0 \Longrightarrow x'\Sigma x > 0$.

(f) $\langle \cdot, \cdot \rangle_{\Sigma^{-1}}$ is bilinear and symmetric. As it is also positive definite from the previous questions it is a scalar product.

(4)

(a) $\langle M - \frac{b}{a}1_d, 1_d \rangle_{\Sigma^{-1}} = \langle M, 1_d \rangle_{\Sigma^{-1}} - \frac{b}{a}\langle 1_d, 1_d \rangle_{\Sigma^{-1}}, = b - \frac{b}{a}a = 0$.

(b) We decompose in an orthogonal basis: $M = (M - \frac{b}{a}1_d) + \frac{b}{a}1_d$,
so $\|M\|_{\Sigma^{-1}}^2 = \|M - \frac{b}{a}1_d\|_{\Sigma^{-1}}^2 + \|\frac{b}{a}1_d\|_{\Sigma^{-1}}^2 = c + \frac{b^2}{a}$.

(5)

(a) $\frac{\partial L_\lambda}{\partial \pi} = M' - 2\lambda\pi'\Sigma - \mu 1_d'$.

(b) We solve the Lagrangian in μ:
$\pi'1_d = 1 \Rightarrow \frac{1}{2\lambda}1_d'\Sigma^{-1}(M - \mu 1_d) = 1 \Rightarrow b - \mu a = 2\lambda \Rightarrow \mu = \frac{b - 2\lambda}{a}$,
so $\pi_\lambda = \frac{1}{2\lambda}\Sigma^{-1}(M - \frac{b - 2\lambda}{a}1_d) = \frac{1}{a}\Sigma^{-1}1_d + \frac{1}{2\lambda}\Sigma^{-1}(M - \frac{b}{a}1_d)$ (standard decomposition).

(c) $\mathbf{E}(\pi_\lambda) = \frac{b}{a} + \frac{c}{2\lambda}$.

(d) $\pi_\lambda'\Sigma\pi = \|\frac{1}{a}1_d + \frac{1}{2\lambda}(M - \frac{b}{a}1_d)\|_{\Sigma^{-1}}^2 = \|\frac{1}{a}1_d\|_{\Sigma^{-1}}^2 + \|\frac{1}{2\lambda}(M - \frac{b}{a}1_d)\|_{\Sigma^{-1}}^2$
$= \frac{1}{a^2}a + \frac{1}{4\lambda^2}c = \frac{1}{a} + \frac{c}{4\lambda^2}$.

(e) $\frac{1}{a}\Sigma^{-1}1_d$. It is the investment portfolio (built uniquely with the risky assets) which has the minimum variance.

(f) The portfolios π_λ are efficient and $\mathcal{F}^+$ is the efficient frontier for risk/return as λ varies between 0 and infinity.

(6)

(a) $\begin{cases} \underset{\pi \in \mathbb{R}^d}{\inf} \ \mathbf{Var}(R^\pi) - \gamma\mathbf{E}(R^\pi) \\ \pi'1_d = 1 \end{cases} = \begin{cases} -\underset{\pi \in \mathbb{R}^d}{\sup} \ -\mathbf{Var}(R^\pi) + \gamma\mathbf{E}(R^\pi) \\ \pi'1_d = 1 \end{cases}$

$= \begin{cases} -\gamma \underset{\pi \in \mathbb{R}^d}{\sup} \ -\frac{1}{\gamma}\mathbf{Var}(R^\pi) + \mathbf{E}(R^\pi) \\ \pi'1_d = 1. \end{cases}$

(b) (P_λ) corresponds to the problem of maximising the expectation for a given level of risk, whereas (P^*_γ) corresponds to the problem of minimising the risk for a given level of expectation. Both methods enable the construction of the efficient frontier.

Exercise

(1)

(a) $R^\pi = \pi' R + (1 - \pi' 1_d) r_0 = \pi'(R - r_0 1_d) + r_0.$
(b) $\mathbf{E}(R^\pi - r_0) = \pi'(M - r_0 1_d).$
(c) $\sigma^\pi = \sqrt{\mathbf{Var}(R^\pi)} = (\pi' \Sigma \pi)^{\frac{1}{2}}.$

(2)

(a) If π^* is a solution of (V) then both π^* and $-\pi^*$ are solutions of (S) (depending on the sign of $\pi^{*\prime}(M - R_0 1_d)$).
(b) $2\mathbf{E}(R^\pi - r_0) \frac{\partial \mathbf{E}(R^\pi - r_0)}{\partial \pi} = 2\pi'(M - r_0 1_d)(M - r_0 1_d)'.$
(c) $2\pi' \Sigma.$
(d) $\frac{\partial}{\partial \pi} \frac{\left(\mathbf{E}(R^\pi - r_0)\right)^2}{\mathbf{Var}(R^\pi)} = 0,$
$\Leftrightarrow \mathbf{Var}(R^\pi) \frac{\partial}{\partial \pi}\left(\mathbf{E}(R^\pi - r_0)\right)^2 - \left(\mathbf{E}(R^\pi - r_0)\right)^2 \frac{\partial}{\partial \pi} \mathbf{Var}(R^\pi) = 0$
$\Leftrightarrow 2\left(\mathbf{E}(R^\pi - r_0)\right)\mathbf{Var}(R^\pi)(M - r_0 1_d)' - 2\left(\mathbf{E}(R^\pi - r_0)\right)^2 \pi' \Sigma = 0$
$\Leftrightarrow \mathbf{Var}(R^\pi)(M - r_0 1_d) - \mathbf{E}(R^\pi - r_0)\Sigma \pi = 0.$

(3)

(a) $M \neq r_0 1_d$ so we can find i such that $r_i \neq r_0$. So,
if $r_i > r_0$ then $\frac{1}{2} r_0 + \frac{1}{2} r_i > r_0,$
if $r_i < r_0$ then $\frac{3}{2} r_0 - \frac{1}{2} r_i > r_0.$
(b) According to 3(a) and 2(b) all portfolios which are solutions are necessarily of this type and conversely all portfolios of this type have the same Sharpe Ratio, equal to $\|M - r_0 1_d\|_{\Sigma^{-1}}.$
(c) We solve α to obtain $\pi' 1_d = 1.$
$\pi' 1_d = 1 \Leftrightarrow \alpha 1'_d \Sigma^{-1}(M - r_0 1_d) = 1 \Leftrightarrow \alpha = \frac{1}{b - r_0 a}.$

10.4   Exam, January 2017

Master M1: Mido 11th January 2017 (Exam: Portfolio Management[4]: Time 2 h)

Notations   We consider an economy with a risk-free asset S^0 of return r_0 and d risky assets $S^1, S^2, \cdots, S^d$ whose returns over $[0, T]$ satisfy $R^i = m^i + \epsilon^i.$

---

[4]Pierre Brugière, University Paris 9 Dauphine.

We write

$$R = M + \epsilon \text{ with } R = \begin{pmatrix} R^1 \\ \vdots \\ R^d \end{pmatrix}, M = \begin{pmatrix} m^1 \\ \vdots \\ m^d \end{pmatrix} \text{ and } \epsilon = \begin{pmatrix} \epsilon^1 \\ \vdots \\ \epsilon^d \end{pmatrix},$$

where M is a vector from $\mathbb{R}^d$ and ϵ is a Gaussian vector of expectation zero and with invertible variance-covariance matrix Σ_d. $\pi' = (\pi^1, \cdots, \pi^d)$ is the vector of allocations at time 0 in the risky assets S^i. π^0 is the allocation at time 0 in the risk-free asset and $\mathbf{1}$ the vector from $\mathbb{R}^d$ whose components are all 1. R^π is the return over $[0, T]$ of the investment portfolio whose allocation in the risky assets is defined by the vector π, m^π is its expected return and σ^π is the standard deviation of its return. $\beta_Q(P)$ is the beta of the investment portfolio P relative to the investment portfolio Q, i.e. $\beta_Q(P) = \frac{\text{Cov}(R_Q, R_P)}{\text{Var}(R_Q)}$.

We denote by R_{T_d} the return of the tangent portfolio T_d from the economy comprising the d risky assets $(S^i)_{i \in [\![1,d]\!]}$ and the risk-free asset S^0. For all investment portfolios P made of the risky assets $(S^i)_{i \in [\![1,d]\!]}$ and the risk free asset S^0 we will have according to the Security Market Line

$$R_P - r_0 = \beta_{T_d}(P)(R_{T_d} - r_0) + \epsilon_P$$

and $m_P - r_0 = \beta_{T_d}(P)(m_{T_d} - r_0)$ (by taking the expectation from the above equation).

Exercise 1   [3pts]

We consider the incomplete table below

Portfolio	$E(R_{P_i})$	$\beta_{T_d}(P_i)$	$\sigma(R_{P_i})$	$\sigma(\epsilon_{P_i})$
P_1	5%	0	0%	?
P_2	10%	1	20%	0%
P_3	?	1.5	30%	0%
P_4	?	?	10%	0%
P_5	?	2	?	30%

where each portfolio P_i follows the Security Market Line equation

$$R_{P_i} - r_0 = \beta_{T_d}(P_i)(R_{T_d} - r_0) + \epsilon_{P_i}.$$

(1)

    (a) From the table above determine r_0, m_{T_d} and σ_{T_d}.   **[1.50pt]**

    (b) Complete the table above.   **[1.50pt]**

Exercise 2   **[3pts]**

We denote by R_λ the return of the investment portfolio built by investing a fraction λ of the initial wealth in the tangent portfolio T_d and the fraction $1 - \lambda$ in the risk-free asset S^0.

(1)

    (a) Prove that $R_\lambda = \lambda R_{T_d} + (1 - \lambda)r_0$.   **[0.50pt]**

    (b) Calculate $m_\lambda = \mathbf{E}(R_\lambda)$ as a function of λ, m_{T_d} and r_0.   **[0.50pt]**

    (c) Calculate $\sigma_\lambda^2 = \mathbf{Var}(R_\lambda)$ as a function of λ and σ_{T_d}.   **[0.50pt]**

    (d) What is the geometric nature of the set   **[0.50pt]**

$$\left\{ \begin{pmatrix} m_\lambda \\ \sigma_\lambda \end{pmatrix}, \lambda \in [0, 1] \right\}?$$

    (e) Calculate $\mathbf{Cov}(R_\lambda, R_{T_d})$ as a function of λ and σ_{T_d}.   **[0.50pt]**

    (f) Calculate $\beta_{T_d}(R_\lambda)$ as a function of λ.   **[0.50pt]**

Exercise 3   **[14pts]**

From now on we add to the economy a new risky asset S^{d+1} of return R^{d+1} and we assume that the completed vector of returns (of dimension $d + 1$) $\begin{pmatrix} R \\ R^{d+1} \end{pmatrix}$ is Gaussian. We denote by Σ_{d+1} the variance-covariance matrix of this vector and $\beta_{T_d}(S^{d+1})$ the beta of S^{d+1} relative to the tangent portfolio T_d (from the economy with only d risky assets). We assume also that there is no possible arbitrage in this larger economy embedding the asset S^{d+1}. We also assume that $m_{d+1} - r_0 > \beta_{T_d}(S^{d+1})(m_{T_d} - r_0)$ (which implies that T_d is not the tangent portfolio T_{d+1} of the bigger economy embedding S^{d+1}).

(1)

    (a) Explain, without doing any calculations, why it is not possible, due to the hypotheses made on m_{d+1}, to build with the assets $(S^i)_{i \in [\![1,d]\!]}$ and the risk free asset S^0 an investment portfolio which replicates the returns of S^{d+1}.   **[0.50pt]**

    (b) Explain, without doing any calculations, why, due to the hypotheses made on m_{d+1}, the matrix Σ_{d+1} is necessarily invertible.   **[1pt]**

    We can admit from now on the result of (1)(b) and the fact that Σ_{d+1} is invertible, and we are now going to search for a vector π from $\mathbb{R}^d$ which minimises $\mathbf{Var}(R^{d+1} - \pi'R)$.

(2)

    (a) Express $\mathbf{Var}(R^{d+1} - \pi'R)$ as a function of $\mathbf{Var}(R^{d+1})$, π, Σ_d and of the vector

        $\mathbf{Cov}(R, R^{d+1})$ whose d components are $\mathbf{Cov}(R^i, R^{d+1})$ for $i \in [\![1, d]\!]$.   **[1pt]**

(b) Show that $\mathbf{Var}(R^{d+1} - \pi'R)$ is minimum for $\pi^* = \Sigma_d^{-1}\mathbf{Cov}(R, R^{d+1})$. **[1pt]**

We denote by P^* the investment portfolio whose allocation in the risky assets $(S^i)_{i \in [\![1,d]\!]}$ is π^* as defined in (2)(b) and whose allocation in S^0 is $1 - \pi^{*\prime}1_d$. We denote by R^* the return of this portfolio P^*. We denote by $P^\perp$ the self-financing portfolio whose allocation in the risky asset S^{d+1} is 1 and whose allocation in the risky assets $(S^i)_{i \in [\![1,d]\!]}$ is $-\pi^*$ and by $R^\perp$ its return.

(3)

   (a) Prove that $R^{d+1} = R^* + R^\perp$. **[0.50pt]**
   (b) Prove that $\forall i \in [\![1, d]\!]$, $\mathbf{Cov}(R^i, R^\perp) = 0$. **[1pt]**
   (c) Prove by using the previous results that **[0.50pt]**

$$\mathbf{Cov}(R^*, R^\perp) = 0 \text{ and } \mathbf{Cov}(R_{T_d}, R^\perp) = 0.$$

   (d) Prove by using the previous results that $\mathbf{Var}(R^\perp) > 0$. **[0.50pt]**
   (e) Prove by using the previous results that $\mathbf{E}(R^\perp) > 0$. **[1pt]**

   We can admit from now on the results from (3) and we define $m_\perp = \mathbf{E}(R^\perp)$ and
$$\sigma_\perp = (\mathbf{Var}(R^\perp))^{\frac{1}{2}}.$$

(4)

   (a) Prove that for all $\alpha \in \mathbb{R}$ it is possible to build an investment portfolio Q_α of return $R_{Q_\alpha} = R_{T_d} + \alpha R^\perp$. **[0.50pt]**
   (b) Using the previous results calculate $\mathbf{E}(R_{Q_\alpha})$ and $\mathbf{Var}(R_{Q_\alpha})$ as a function of $\alpha, m_\perp, m_{T_d}, r_0, \sigma_{T_d}$ and $\sigma_\perp$. **[0.50pt]**
   (c) Let α^* be the solution of: $\frac{\partial}{\partial \alpha}\left(\frac{\mathbf{E}(R_{Q_\alpha})-r_0}{\mathbf{Var}(R_{Q_\alpha})^{\frac{1}{2}}}\right) = 0$ show that: $\alpha^* =$

$\frac{m_\perp}{m_{T_d} - r_0}\frac{\sigma_{T_d}^2}{\sigma_\perp^2}$.

   (d) For all investment portfolios Q_α calculate $\beta_{Q_{\alpha^*}}(Q_\alpha)$ as a function of $\alpha, \alpha^*, m_\perp, m_{T_d}$ and r_0. **[1pt]**
   (e) For all investment portfolios Q_α prove through calculations that **[1pt]**

$$m_{Q_\alpha} - r_0 = \beta_{Q_{\alpha^*}}(Q_\alpha)(m_{Q_{\alpha^*}} - r_0).$$

   (f) For all investment portfolios Z built with the risky assets $(S^i)_{i \in [\![1,d]\!]}$ and the risk-free asset S^0 calculate $\beta_{Q_{\alpha^*}}(Z)$ as a function of $m_Z, r_0, m_{T_d}, \alpha^*, m_\perp$. **[1pt]**
   (g) For all investment portfolios Z built with the risky assets $(S^i)_{i \in [\![1,d]\!]}$ and the risk-free asset S^0 prove through calculations that $m_Z - r_0 = \beta_{Q_{\alpha^*}}(Z)(m_{Q_{\alpha^*}} - r_0)$. **[1pt]**
   (h) What interesting things can you say (and which characterises it) about the portfolio Q_{α^*}? **[1pt]**

10.4.1 Solutions: Exam, January 2017

Exercise 1

(1)

  (a) $\sigma(R_{P_1}) = 0$, so P_1 is the risk free asset, and so $r_0 = 5\%$.
  $\sigma(\epsilon_{P_2}) = 0$ and $\beta_{T_d}(P_2) = 1$, so P_2 is the tangent portfolio and $m_{T_d} = 10\%$
  and $\sigma_{T_d} = 20\%$.

  (b) $\sigma(R_{P_1}) = 0 \Longrightarrow \sigma(\epsilon_{P_1}) = 0$.
  $\mathbf{E}(R_{P_3}) = 5\% + 1.5(10\% - 5\%) = 12.5\%$.
  $\sigma(R_{P_4}) = 0 \Longrightarrow \beta_{T_d}(P_2) = 0.5 \Longrightarrow \mathbf{E}(R_{P_4}) = 5\% + 0.5(10\% - 5\%) = 7.5\%$.
  $\beta_{T_d}(P_5) = 2 \Longrightarrow \mathbf{E}(R_{P_5}) = 5\% + 2(10\% - 5\%) = 15\%$.
  $\beta_{T_d}(P5) = 2 \Longrightarrow \sigma(R_{P_i})^2 = 4 \times (20\%)^2 + (30\%)^2 \Longrightarrow \sigma(R_{P_i}) = 50\%$.

Exercise 2

(1)

  (a) $R_\lambda = \begin{pmatrix} 1 - \lambda \\ \lambda \end{pmatrix}' \begin{pmatrix} r_0 \\ R_{T_d} \end{pmatrix}$.

  (b) $m_\lambda = \mathbf{E}[R_\lambda] = \lambda m_{T_d} + (1 - \lambda) r_0$.

  (c) $\sigma_\lambda^2 = \lambda^2 \sigma_{T_d}^2$.

  (d) $\begin{pmatrix} m_\lambda \\ \sigma_\lambda \end{pmatrix} = \lambda \begin{pmatrix} m_{T_d} \\ \sigma_{T_d} \end{pmatrix} + (1 - \lambda) \begin{pmatrix} r_0 \\ 0 \end{pmatrix}$, the parametric representation of a segment.

  (e) $\mathbf{Cov}(R_\lambda, R_{T_d}) = \mathbf{Cov}(\lambda R_{T_d}, R_{T_d}) = \lambda \sigma_{T_d}^2$.

  (f) $\beta_{T_d}(P_\lambda) = \mathbf{Cov}(R_\lambda, R_{T_d}) \frac{1}{\sigma_{T_d}^2} = \lambda$.

Exercise 3

(1)

  (a) All investment portfolios built with the assets $(S^i)_{i \in [\![1,d]\!]}$ and S^0 must satisfy the SML, so if the SML is not satisfied for S^{d+1} it means S^{d+1} cannot be replicated by these assets.

  (b) If Σ_{d+1} was not invertible we could find a combination of the returns of the $d + 1$ risky assets with a fixed return. This would imply that either there is a self-financing portfolio or an investment portfolio built with the $d + 1$ risky assets which is without risk. In both cases this combination would have a non-negative coefficient on R_{d+1} as the hypothesis that Σ_d is invertible means it is not possible to build such a portfolio with the assets $(S^i)_{i \in [\![1,d]\!]}$ only. Therefore if Σ_{d+1} was not invertible S^{d+1} could be replicated by the assets $(S^i)_{i \in [\![1,d]\!]}$ and S^0, which is impossible according to (a).

(2)

    (a) $\mathbf{Var}(R^{d+1} - \pi'R) = \mathbf{Var}(R^{d+1} - \pi'R, R^{d+1} - \pi'R) = \mathbf{Var}(R^{d+1}) - 2\pi'\mathbf{Cov}(R, R^{d+1}) + \pi'\Sigma_d\pi$.

    (b) Let $\phi(\pi) = \mathbf{Var}(R^{d+1} - \pi'R)$. We have
$$\frac{\partial\phi}{\partial\pi} = 0 \Longrightarrow -2\mathbf{Cov}(R, R^{d+1}) + 2\Sigma_d\pi = 0 \Longrightarrow \pi = \Sigma_d^{-1}\mathbf{Cov}(R, R^{d+1}).$$

(3)

    (a) $R^* = \pi^{*\prime}R + (1 - \pi^{*\prime}\mathbf{1})r_0$ and $R^{\perp} = R^{d+1} - \pi^{*\prime}R - (1 - \pi^{*\prime}\mathbf{1})r_0$.

    (b) Let's prove that $\mathbf{Cov}(R, R^{\perp}) = 0$, which will prove that for each component $\mathbf{Cov}(R^i, R^{\perp}) = 0$.
$$\mathbf{Cov}(R, R^{\perp}) = \mathbf{Cov}(R, R^{d+1} - R^*) = \mathbf{Cov}(R, R^{d+1}) - \mathbf{Cov}(R, [\Sigma_d^{-1}\mathbf{Cov}(R, R^{d+1})]'R)$$
$$= \mathbf{Cov}(R, R^{d+1}) - \mathbf{Cov}(R, R)\Sigma_d^{-1}\mathbf{Cov}(R, R^{d+1})$$
$$= \mathbf{Cov}(R, R^{d+1}) - \Sigma_d\Sigma_d^{-1}\mathbf{Cov}(R, R^{d+1}) = 0.$$

    (c) $\mathbf{Cov}(R^*, R^{\perp}) = \mathbf{Cov}(\pi^{*\prime}R, R^{\perp}) = \pi^{*\prime}\mathbf{Cov}(R, R^{\perp}) = \pi^{*\prime} \times 0$ (according to 3(b). In the same way the risky components of the R_{T_d} are the R_i which satisfy $\mathbf{Cov}(R^i, R^{\perp}) = 0$, so by linearity $\mathbf{Cov}(R_{T_d}, R^{\perp}) = 0$.

    (d) $\mathbf{Var}(R^{\perp}) = \mathbf{Var}(R^{d+1} - \pi^{*\prime}R) = \begin{pmatrix} -\pi^* \\ 1 \end{pmatrix}' \Sigma_{d+1} \begin{pmatrix} -\pi^* \\ 1 \end{pmatrix}$.

    As Σ_{d+1} is positive definite (by assumption) the quadratic product of a non-zero vector is > 0.

    (e) According to the hypothesis $m_{d+1} - r_0 > \beta_{T_d}(S^{d+1})(m_{T_d} - r_0)$.
$$\beta_{T_d}(S^{d+1}) = \mathbf{Cov}(R^{d+1}, R^{T_d})\frac{1}{\sigma_{T_d}^2} = \mathbf{Cov}(R^*, R^{T_d})\frac{1}{\sigma_{T_d}^2},$$
so $\beta_{T_d}(S^{d+1})(m_{T_d} - r_0) = \beta_{T_d}(P^*)(m_{T_d} - r_0) = (m_{P^*} - r_0)$
so, $m_{d+1} - r_0 > \beta_{T_d}(S^{d+1})(m_{T_d} - r_0) \Longrightarrow m_{d+1} - r_0 > m_{P^*} - r_0 \Longrightarrow m_{\perp} > 0$.

(4)

    (a) When allocating 1 to the tangent portfolio T_d and α to the self-financing portfolio $P^{\perp}$ we obtain an investment portfolio whose return is $R_{T_d} + \alpha R^{\perp}$.

    (b) $\mathbf{E}(R_{Q_\alpha}) = m_{T_d} + \alpha m_{\perp}$.
$\mathbf{Var}(R_{Q_\alpha}) = \mathbf{Var}(R_{T_d}) + \alpha^2\mathbf{Var}(R^{\perp})$ (the covariance term is zero according to 3(c). So, $\mathbf{Var}(R_{Q_\alpha}) = \sigma_{M_{T_d}}^2 + \alpha^2\sigma_{\perp}^2$.

    (c) $m_{\perp}\sigma(R_{Q_\alpha}) - \mathbf{E}(R_{Q_\alpha} - r_0)\alpha\sigma_{\perp}^2\frac{1}{\sigma(R_{Q_\alpha})} = 0$
$$\Longleftrightarrow m_{\perp}(\sigma_{T_d}^2 + \alpha^2\sigma_{\perp}^2) - [(m_{T_d} - r_0) + \alpha m_{\perp}]\alpha\sigma_{\perp}^2 = 0$$
$$\Longleftrightarrow m_{\perp}\sigma_{T_d}^2 - (m_{T_d} - r_0)\alpha\sigma_{\perp}^2 = 0.$$

    (d) $\beta_{Q_{\alpha^*}}(Q_\alpha) = \frac{\mathbf{Cov}(R_{Q_{\alpha^*}}, R_{Q_\alpha})}{\mathbf{Var}(R_{Q_{\alpha^*}})}$,
$\mathbf{Cov}(R_{Q_{\alpha^*}}, R_{Q_\alpha}) = \mathbf{Cov}(R_{T_d} + \alpha^* R^{\perp}, R_{T_d} + \alpha R^{\perp}), = \sigma_{T_d}^2 + \alpha\alpha^*\sigma_{\perp}^2$,
$\mathbf{Var}(R_{Q_{\alpha^*}}) = \sigma_{T_d}^2 + \alpha^{*2}\sigma_{\perp}^2$,
$$\frac{\mathbf{Cov}(R_{Q_{\alpha^*}}, R_{Q_\alpha})}{\mathbf{Var}(R_{Q_{\alpha^*}})} = \frac{1 + \frac{\alpha\alpha^*\sigma_{\perp}^2}{\sigma_{T_d}^2}}{1 + \frac{\alpha^{*2}\sigma_{\perp}^2}{\sigma_{T_d}^2}} = \frac{1 + \alpha\frac{m_{\perp}}{m_{T_d} - r_0}}{1 + \alpha^*\frac{m_{\perp}}{m_{T_d} - r_0}}.$$

(e) $(m_{Q_{\alpha*}} - r_0) = (m_{T_d} + \alpha^* m_\perp - r_0) = (m_{T_d} - r_0)(1 + \frac{\alpha^* m_\perp}{m_{T_d} - r_0})$,

so $\beta_{Q_{\alpha*}}(Q_\alpha)(m_{Q_{\alpha*}} - r_0) = \frac{1 + \alpha \frac{m_\perp}{m_{T_d} - r_0}}{1 + \alpha^* \frac{m_\perp}{m_{T_d} - r_0}}(m_{T_d} - r_0)(1 + \frac{\alpha^* m_\perp}{m_{T_d} - r_0})$

$= m_{T_d} - r_0 + \alpha m_\perp = m_{Q_\alpha} - r_0$.

(f) $\mathbf{Cov}(R_Z, R_{Q^*}) = \mathbf{Cov}(R_Z, R_{T_d}) = \frac{m_Z - r_0}{m_{T_d} - r_0}\sigma_{T_d}^2$ (as R^Z satisfies the SML),

so $\beta_{Q_{\alpha*}}(Z) = \frac{m_Z - r_0}{m_{T_d} - r_0}\frac{\sigma_{T_d}^2}{\sigma_{T_d}^2 + \alpha^{*2}\sigma_\perp^2} = \frac{m_Z - r_0}{m_{T_d} - r_0}\frac{1}{1 + \frac{\alpha^{*2}\sigma_\perp^2}{\sigma_{T_d}^2}} = \frac{m_Z - r_0}{m_{T_d} - r_0}\frac{1}{1 + \alpha^* \frac{m_\perp}{m_{T_d} - r_0}}$.

(g) $m_{Q_{\alpha*}} - r_0 = (m_{T_d} - r_0)(1 + \frac{\alpha^* m_\perp}{m_{T_d} - r_0})$.

So, $\beta_{Q_{\alpha*}}(Z)(m_{Q_{\alpha*}} - r_0) = \frac{m_Z - r_0}{m_{T_d} - r_0}\frac{1}{1 + \alpha^* \frac{m_\perp}{m_{T_d} - r_0}}(m_{T_d} - r_0)(1 + \frac{\alpha^* m_\perp}{m_{T_d} - r_0}) =$

$m_Z - r_0$.

(h) It is the Tangent Portfolio for the bigger economy, i.e. $Q_{\alpha*} = T_{d+1}$.

10.5 Midterm Exam, November 2017

Master M1: Mido 2nd November 2017 (Midterm Exam: Portfolio Management[5]: Time 2 h)

Notations We consider d risky assets $S^1, S^2, ...S^d$, with returns on $[0, T]$ satisfying $R^i = m^i + \epsilon^i$.
  We define:

- $R = M + \epsilon$ with $R = \begin{pmatrix} R^1 \\ \vdots \\ R^d \end{pmatrix}$, $M = \begin{pmatrix} m^1 \\ \vdots \\ m^d \end{pmatrix}$ and $\epsilon = \begin{pmatrix} \epsilon^1 \\ \vdots \\ \epsilon^d \end{pmatrix}$,

where M is a vector from $\mathbb{R}^d$ and ϵ is a Gaussian vector of expectation zero and with invertible variance-covariance matrix Σ.

- $\pi' = (\pi^1, \cdots, \pi^d)$ the vector of the allocations at time zero in the assets S^i.
- R^π the return of the portfolio π on $[0, T]$, m_π the expectation of its return and σ_π the standard deviation of its return.
- 1_d the vector of $\mathbb{R}^d$ with all its components equal to 1.
- $a = 1_d'\Sigma^{-1}1_d$ and $b = 1_d'\Sigma^{-1}M$.

For a function $L(\pi)$ we denote by $\frac{\partial L}{\partial \pi}$ the row vector $\left(\frac{\partial L}{\partial \pi^1}, \cdots, \frac{\partial L}{\partial \pi^d}\right)$.

---

[5]Pierre Brugière, University Paris 9 Dauphine.

Exercise 1 [4pts]

Answer the following questions without giving justifications:

(1) Express $\mathbf{E}(R^\pi)$ as a function of π and M. [0.25pt]

(2) Express $\mathbf{Var}(R^\pi)$ as a function of π and Σ. [0.25pt]

(3) Express $\mathbf{Cov}(R^{\pi_1}, R^{\pi_2})$ as a function of π_1, π_2 and Σ. [0.50pt]

  Let X and Y be two random vectors of $\mathbb{R}^k$ and $\mathbb{R}^l$, A a matrix of $\mathbb{R}^{m \times k}$ and B a matrix of $\mathbb{R}^{n \times l}$.

(4) Express $\mathbf{Cov}(X, Y)$ as a function of $\mathbf{E}(XY')$, $\mathbf{E}(X'Y)$, $\mathbf{E}(X)$ and $\mathbf{E}(Y)$. [0.50pt]

(5) Express $\mathbf{Cov}(AX, BY)$ as a function of A, B and $\mathbf{Cov}(X, Y)$. [0.50pt]

(6) Express $\mathbf{Cov}(X, Y)$ as a function of $\mathbf{Cov}(Y, X)$. [0.50pt]

(7) Express $\frac{\partial f}{\partial \pi}$ when $f(\pi) = \pi' \Sigma \pi + \alpha_1 \pi' M + \alpha_2$. [0.50pt]

  We consider a strategy (x_0, π).

(8) What is the value of $1'_d \pi$ for an investment strategy? [0.25pt]

(9) What is the value of $1'_d \pi$ for a self-financing strategy? [0.25pt]

(10) What does x_0 represent for an investment strategy? [0.25pt]

(11) What does x_0 represent for a self-financing strategy? [0.25pt]

Exercise 2 [6pts]

We assume that $M \neq 0$ and let $\pi_M = \frac{1}{b}\Sigma^{-1}M$ and R_M be the return of the portfolio (x_0, π_M).

(1) Prove that π_M is the allocation of an investment portfolio. [0.25pt]

  We consider an investment portfolio (x_0, π_P) whose return is denoted R_P.

(2) Show that $\mathbf{Cov}(R_P, R_M) = 0 \implies \mathbf{E}[R_P] = 0$. [0.25pt]

  We consider the problem $\inf_{\alpha \in \mathbb{R}} \mathbf{Var}(R_P - \alpha R_M)$.

(3) Show that the inf is reached for a unique value of α, denoted $\alpha_M(P)$. [0.25pt]

(4) Express $\alpha_M(P)$ as a function of $\mathbf{Cov}(R_M, R_P)$, $\mathbf{Var}(R_M)$ and $\mathbf{Var}(R_P)$. [0.25pt]

(5) Calculate the numerical value of $\mathbf{E}(R_P - \alpha_M(P)R_M)$. [1pt]

(6) Express $\mathbf{Var}(R_P - \alpha_M(P)R_M)$ as a function of $\mathbf{Var}(R_P)$ and of the correlation between R_P and R_M, denoted $\rho_{P,M}$. [1pt]

(7) Show, based on the previous results, that for any investment portfolio P of return R_P we have $R_P = \alpha_{P,M} R_M + \epsilon_{P,M}$ with $\alpha_{P,M} \in \mathbb{R}$ and

  (i) $\mathbf{E}(\epsilon_{P,M}) = 0$,

  (ii) $\mathbf{Var}(\epsilon_{P,M}) \leq \mathbf{Var}(R_P)$. [0.50pt]

(8) Show that if an investment portfolio (x_0, π_Q) satisfies

$$\mathbf{Cov}(R_P, R_Q) = 0 \implies \mathbf{E}(R_P) = 0,$$

then $\pi_Q = \pi_M$. [1.5pt]

(9) What do (7) and (8) imply in terms of "risk remuneration" for an investment portfolio. [1pt]

Exercise 3 [5pts]

We denote by R_Z the return of an asset which may or may not belong to the universe of portfolios that can be built with the risky assets $S^1, S^2, \cdots S^d$. We want to solve

$$
\begin{cases}
\inf_{\pi \in \mathbb{R}^d} \mathbf{Var}(R_Z - \pi' R) \\
\qquad \pi' 1_d = 1
\end{cases}
$$

(1) Write the expression of the Lagrangian $L(\pi, \lambda)$.   [1.00pt]
(2) Calculate $\frac{\partial L}{\partial \pi}$.   [0.50pt]
(3) Justify that the inf is reached for a solution π of $\frac{\partial L}{\partial \pi} = 0$.   [0.50pt]
(4) Show that the solution of the problem can be expressed as $\pi = \pi_a + \Sigma^{-1}[U_Z - \pi_a' U_Z 1_d]$. What are the expressions of U_Z and π_a?   [2.00pt]
(5) To which situation corresponds the case $U_Z = 0$ and how can the solution of the optimisation problem be found directly in this case?   [1.00pt]

Exercise 4 [5pts]

We recall that if $\mathcal{P}$ is the set of all investment portfolios that can be built, then

$$
\{(\sigma_\pi, m_\pi)', \pi \in \mathcal{P}\}
$$

is delimited by the set $\{(\sigma_\pi, m_\pi)', \pi \in \mathcal{F}\}$, where $\mathcal{F}$ is the set of investment portfolios of the form

$$
\frac{1}{a}\Sigma^{-1}1_d + \lambda\Sigma^{-1}(M - \frac{b}{a}1_d)
$$

for $\lambda \in \mathbb{R}$.

(1) Show that $(\alpha \in \mathbb{R}, \pi_1 \in \mathcal{F}$ and $\pi_2 \in \mathcal{F}) \Longrightarrow \alpha\pi_1 + (1 - \alpha)\pi_2 \in \mathcal{F}$.   [1pt]
(2) Show that $(\alpha \in \mathbb{R} \setminus \{0\}, \pi_1 \in \mathcal{P} \setminus \mathcal{F}$ and $\pi_2 \in \mathcal{F}) \Longrightarrow \alpha\pi_1 + (1 - \alpha)\pi_2 \in \mathcal{P} \setminus \mathcal{F}$.   [1pt]
    We assume in all what follows that $M \neq \frac{b}{a}1_d$, $\pi_P, \pi_Q \in \mathcal{F}$ with $m_P \neq m_Q$ and that $\pi_R \in \mathcal{P} \setminus \mathcal{F}$ with $m_R \neq m_P$.
(3) Show that $\forall \pi \in \mathcal{F} \exists \alpha \in \mathbb{R}, \pi = \alpha\pi_P + (1 - \alpha)\pi_Q$.   [1pt]
(4) Show that $\forall(\sigma, m)'$ to the right of $\mathcal{F}$ we can find $\alpha_1, \alpha_2 \in \mathbb{R}$ such that the investment portfolio of allocation $\alpha_1\pi_P + \alpha_2\pi_Q + (1 - \alpha_1 - \alpha_2)\pi_R$ has an expected return of m and a standard deviation of the return of σ.   [2pt]

10.5.1 Solutions: Midterm Exam, November 2017

Exercise 1

(1) $\pi'M$.
(2) $\pi'\Sigma\pi$.
(3) $\pi_1'\Sigma\pi_2$.
(4) $\mathbf{E}(XY') - \mathbf{E}(X)\mathbf{E}(Y)'$.
(5) $A\mathbf{Cov}(X, Y)B'$.
(6) $\mathbf{Cov}(X, Y) = \mathbf{Cov}(Y, X)'$.
(7) $\frac{\partial f}{\partial \pi} = 2\pi'\Sigma + \alpha_1 M'$.
(8) 1.
(9) 0.
(10) The initial wealth.
(11) The notional.

Exercise 2

(1) $1_d'\pi_M = \frac{b}{b} = 1$.
(2) $\mathbf{Cov}(R_P, R_M) = \pi_P'\Sigma\frac{1}{b}\Sigma^{-1}M = \frac{1}{b}\pi_P'M = \frac{1}{b}\mathbf{E}(R_P)$.
(3) $\mathbf{Var}(R_P - \alpha R_M) = \mathbf{Var}(R_P) + \alpha^2\mathbf{Var}(R_M) - 2\alpha\mathbf{Cov}(R_P, R_M)$, a quadratic function of α with the term in α^2 positive.
(4) The derivative is zero for $2\alpha\mathbf{Var}(R_M) - 2\mathbf{Cov}(R_P, R_M) = 0$, so $\alpha_M(P) = \frac{\mathbf{Cov}(R_P, R_M)}{\mathbf{Var}(R_M)}$.
(5) $\mathbf{E}(R_P - \alpha_M(P)R_M) = \Pi_P'M - \frac{\mathbf{Cov}(R_M, R_P)}{\mathbf{Var}(R_M)}\mathbf{E}(R_M)$
$= \Pi_P'M - \frac{1}{b}\Pi_P'\Sigma\Sigma^{-1}M\frac{\mathbf{E}(R_M)}{\mathbf{Var}(R_M)} = \Pi_P'M - \frac{1}{b}\Pi_P'M\frac{\mathbf{E}(R_M)}{\mathbf{Var}(R_M)}$,
but $\mathbf{E}(R_M) = \frac{1}{b}M'\Sigma^{-1}M$ and $\mathbf{Var}(R_M) = \frac{1}{b^2}M'\Sigma^{-1}\Sigma\Sigma^{-1}M = \frac{1}{b}\mathbf{E}(R_M)$,
so $\mathbf{E}(R_P - \alpha_M(P)R_M) = 0$.
(6) $\mathbf{Var}(R_P - \alpha_M(P)R_M) = \sigma_P^2 - 2\alpha_M(P)\mathbf{Cov}(R_P, R_M) + \alpha_M(P)^2\sigma_M^2$
$= \sigma_P^2 - 2\frac{\mathbf{Cov}(R_P, R_M)^2}{\mathbf{Var}(R_M)} + \frac{\mathbf{Cov}(R_P, R_M)^2}{\mathbf{Var}(R_M)^2}\sigma_M^2$
$= \sigma_P^2 - 2\rho_{P,M}^2\frac{\sigma_P^2\sigma_M^2}{\sigma_M^2} + \rho_{P,M}^2\frac{\sigma_P^2\sigma_M^2}{\sigma_M^4}\sigma_M^2$
$= (1 - \rho_{P,M}^2)\mathbf{Var}(R_P)$.
(7) If we take $\alpha_{P,M} = \alpha_M(P)$ as before, then $\mathbf{E}(\epsilon_{P,M}) = 0$ as $\mathbf{E}(R_P - \alpha_M(P)R_M) = 0$ according to (5) and according to (6)

$$\mathbf{Var}(\epsilon_{P,M}) = \mathbf{Var}(R_P - \alpha_M(P)R_M) = (1 - \rho_{P,M})^2\mathbf{Var}(R_P) \leq \mathbf{Var}(R_P).$$

(8) If R_Q has the property then

$$\{x \in \mathbb{R}^d, x'\Sigma\pi_Q = 0\} \subset \{x \in \mathbb{R}^d, x'M = 0\}$$

$$\implies \left(\text{Vect}(\Sigma\pi_Q)\right)^\perp \subset \left(\text{Vect}(M)\right)^\perp$$

$$\Longrightarrow \text{Vect}(M) \subset \text{Vect}(\Sigma \pi_Q)$$

$$\Longrightarrow \exists \lambda \in \mathbb{R}, M = \lambda \Sigma \pi_Q \text{ (and } \lambda \neq 0 \text{ as } M \neq 0)$$

$$\Longrightarrow \pi_Q = \frac{1}{\lambda} \Sigma^{-1} M.$$

(9) Only the risk correlated to R_M is remunerated and R_M is defined in a unique way by this property.

Exercise 3

(1) $L(\pi, \lambda) = \text{Var}(R_Z) - 2\text{Cov}(R_Z, \pi'R) + \text{Var}(\pi'R) - \lambda(\pi'1_d - 1)$
$= \text{Var}(R_Z) - 2\text{Cov}(R_Z, R)\pi + \pi'\Sigma\pi - \lambda(\pi'1_d - 1).$
(2) $\frac{\partial L}{\partial \pi} = -2\text{Cov}(R_Z, R) + 2\pi'\Sigma - \lambda 1_d'.$
(3) Σ is positive definite so this is a classical result for strictly convex functions.
(4) $\frac{\partial L}{\partial \pi} = 0 \Longrightarrow -2\text{Cov}(R_Z, R) + 2\pi'\Sigma - \lambda 1_d' = 0$
$\Longrightarrow \Sigma'\pi = \text{Cov}(R, R_Z) + \frac{1}{2}\lambda 1_d$
$\Longrightarrow \pi = \Sigma^{-1}\text{Cov}(R, R_Z) + \frac{1}{2}\lambda\Sigma^{-1}1_d.$
The constraint implies $1_d'\Sigma^{-1}\text{Cov}(R, R_Z) + \frac{1}{2}\lambda 1_d'\Sigma^{-1}1_d = 1$,
so $\frac{1}{2}\lambda = \frac{1}{a} - \frac{1}{a}(\Sigma^{-1}1_d)'\text{Cov}(R, R_Z).$
If we define $\pi_a = \frac{1}{a}\Sigma^{-1}1_d$ and $U_Z = \text{Cov}(R, R_Z)$ we get

$$\pi = \Sigma^{-1}U_Z + (\frac{1}{a} - \pi_a'U_Z)\Sigma^{-1}1_d = \pi_a + \Sigma^{-1}(U_Z - \pi_a'U_Z 1_d).$$

(5) $U_Z = 0 \Longleftrightarrow R_Z$ is not correlated to any R_i so $\text{Cov}(R_Z, R_P) = 0$ for all investment portfolios R_P, and so $\text{Var}(R_Z - R_P) = \text{Var}(R_Z) + \text{Var}(R_P)$, hence the minimum is reached when $\text{Var}(R_P)$ is at a minimum and therefore for π_a, which is the solution obtained in (4) when replacing U_Z by zero.

Exercise 4

(1) $\alpha\pi_1 + (1 - \alpha)\pi_2$
$= \alpha[\frac{1}{a}\Sigma^{-1}1_d + \lambda_1\Sigma^{-1}(M - \frac{b}{a}1_d)] + (1 - \alpha)[\frac{1}{a}\Sigma^{-1}1_d + \lambda_2\Sigma^{-1}(M - \frac{b}{a}1_d)]$
$= \frac{1}{a}\Sigma^{-1}1_d + (\alpha\lambda_1 + (1 - \alpha)\lambda_2)\Sigma^{-1}(M - \frac{b}{a}1_d).$
(2) By contradiction

$$\alpha\pi_1 + (1 - \alpha)\pi_2 = \pi \in \mathcal{F} \Longrightarrow \pi_1 = \frac{1}{\alpha}\pi - \frac{1 - \alpha}{\alpha}\pi_2$$

but $\left(\frac{1}{\alpha} - \frac{1-\alpha}{\alpha} = 1 \text{ and } \pi \text{ and } \pi_2 \in \mathcal{F}\right) \Longrightarrow \pi_1 \in \mathcal{F}$ according to (1) and then this leads to a contradiction.
(3) α such that $\lambda_\pi = \alpha\lambda_{\pi_P} + (1 - \alpha)\lambda_{\pi_Q}$ works.

(4) Let α satisfy $\alpha m_P + (1 - \alpha)m_Q = m$ and Z_1 be the investment portfolio of allocation $\alpha \pi_P + (1 - \alpha)\pi_Q$. Let β satisfy $\beta m_P + (1 - \beta)m_R = m$ and Z_2 be the investment portfolio of allocation $\beta \pi_P + (1 - \beta)\pi_R$. If we now consider all the investment portfolios $\{\gamma \pi_{Z_2} + (1 - \gamma)\pi_{Z_1}, \gamma \in \mathbb{R}\}$ then for all portfolios of return R_γ in this set we get $\mathbf{E}(R_\gamma) = m$ and

$$\mathbf{Var}(R_\gamma) = \|\gamma (\pi_{Z_2} - \pi_{Z_1}) + \pi_{Z_1}\|_\Sigma,$$

which takes all values between $\|\pi_{Z_1}\|_\Sigma = \mathbf{Var}(R_{Z_1})$ (for $\gamma = 0$) and $+\infty$. We can therefore find γ^* such that $\mathbf{Var}(R_{\gamma^*}) = \sigma^2$ and so the investment portfolio is a solution of $(1 - \gamma^*)[\alpha \pi_P + (1 - \alpha)\pi_Q] + \gamma^*[\beta \pi_P + (1 - \beta)\pi_R]$.

10.6 Exam, January 2018

Master M1: Mido 15th January 2018 (Exam: Portfolio Management[6]: Time 2 h)

Notations We consider d risky assets $S_1, S_2, ...S_d$, whose returns over $[0, T]$ satisfy $R^i = m^i + \epsilon^i$. In vector form, we write

$$R = M + \epsilon \text{ with } R = \begin{pmatrix} R^1 \\ \vdots \\ R^d \end{pmatrix}, \ M = \begin{pmatrix} m^1 \\ \vdots \\ m^d \end{pmatrix} \text{ and } \epsilon = \begin{pmatrix} \epsilon^1 \\ \vdots \\ \epsilon^d \end{pmatrix},$$

where M is a vector of $\mathbb{R}^d$, ϵ is a random vector of expectation zero and with invertible matrix of variance-covariance Σ. We define:

$$\pi = \begin{pmatrix} \pi^1 \\ \vdots \\ \pi^d \end{pmatrix} \text{ an allocation, at time 0, in the risky assets } S_i,$$

- R_π the return of the portfolio π over $[0, T]$,
- 1_d the vector of $\mathbb{R}^d$ with all components equal to 1,
- $a = 1'_d \Sigma^{-1} 1_d$ and $b = 1'_d \Sigma^{-1} M$,
- r_0 the risk-free rate of the risk-free asset S^0 and we assume that $r_0 \neq \frac{b}{a}$.

Exercise 1 [4pts]
Give without any justification the answers to the following questions.

---

[6]Pierre Brugière, University Paris 9 Dauphine.

(1) Express $\mathbf{Cov}(R_{\pi_1}, R_{\pi_2})$ as a function of π_1, π_2 and Σ.   **[0.25pt]**

(2) If P is an investment portfolio and π_P is its allocation in the risky assets, express its allocation π_P^0 in the risk-free asset as a function of π_P and 1_d.   **[0.25pt]**

(3) If Q is a self-financing portfolio and π_Q is its allocation in the risky assets, express its allocation π_Q^0 in the risk-free asset as a function of π_Q and 1_d.   **[0.25pt]**

(4) What is special about the portfolio $\pi_a = \frac{1}{a}\Sigma^{-1}1_d$?   **[0.25pt]**

   (a) Express $\mathbf{E}(R_{\pi_a})$ as a function of a and b.   **[0.25pt]**

   (b) Express $\mathbf{Var}(R_{\pi_a})$ as a function of a and b.   **[0.25pt]**

(5) If an investment portfolio is of the form $\pi = \alpha\Sigma^{-1}(M - r_0 1_d)$:

   (a) What is the value of α?   **[0.25pt]**

   (b) What is the name of this investment portfolio?   **[0.25pt]**

(6) From the Security Market Line theorem, which says that for any investment portfolio P, $R_P - r_0 = \mathbf{Cov}(R_P, R_T)(R_T - r_0) + \epsilon_P$, deduce a similar equation satisfied by the return R_Q of a self-financing portfolio.   **[1pt]**

(7) In a factor model $R = A + BF + \mathcal{E}$ with $\mathbf{Cov}(F, \mathcal{E}) = 0$ what is the expression of $\mathbf{Var}(R)$ as a function of B, $\mathbf{Var}(F)$ and $\mathbf{Var}(\mathcal{E})$?   **[0.5pt]**

(8) If Σ is a matrix of $\mathbb{R}^d \times \mathbb{R}^d$ which is symmetric positive definite with eigenvalues $(\lambda^i)_{i\in[\![1,d]\!]}$ and a corresponding basis of orthonormal eigenvectors $(u_i)_{i\in[\![1,d]\!]}$ express Σ in terms of the λ^i and u_i.   **[0.5pt]**

Exercise 2   **[3pts]**

Consider the following table:

Portfolio	E(Return)	$\beta_T(P_i)$	$\sigma(R_{P_i})$	$\sigma(\epsilon_{P_i})$
P_1	7%	0.5	5%	0%
P_2	?	1	$\sqrt{\frac{2}{10}}$	10%
P_3	16%	2	20%	0%
P_4	6%	?	?	10%

Recall the Security Market Line equation: $R_{P_i} - r_0 = \beta_T(P_i)(R_T - r_0) + \epsilon_{P_i}$.

(1) From the table deduce r_0, m_T and σ_T.   **[1.50pt]**

(2) Complete the table.   **[1.50pt]**

Exercise 3   **[3pts]**

We consider the factor model $R = \lambda^0 1_3 + BF + \mathcal{E}$, with

$$R = \begin{pmatrix} R^1 \\ R^2 \\ R^3 \end{pmatrix}, \quad B = \begin{pmatrix} 1 & 1 \\ 0 & 2 \\ 2 & 0 \end{pmatrix}, \quad F = \begin{pmatrix} f^1 \\ f^2 \end{pmatrix} \text{ and } \mathcal{E} = \begin{pmatrix} \epsilon^1 \\ \epsilon^2 \\ \epsilon^3 \end{pmatrix}$$

and the assumptions: $\mathbf{Var}(F)$ invertible, $\mathbf{E}(\mathcal{E}) = 0$ and $\mathbf{Cov}(F, \mathcal{E}) = 0$.

(1) Exhibit a self-financing portfolio π_S such that $\mathbf{Cov}(R_{\pi_S}, F) = 0$ and calculate for this portfolio $\mathbf{E}(R_{\pi_S})$.   [1pt]

(2) Is it possible to find in this model an investment portfolio π such that $\mathbf{Cov}(R_\pi, F) = 0$?   [1pt]

(3) We assume that $\mathbf{Var}(F) = 0.01 \times \mathrm{Id}_{\mathbb{R}^2}$ and that $\mathbf{Var}(\mathcal{E}) = 0.01 \times \mathrm{Id}_{\mathbb{R}^3}$. Calculate the correlation between R^1 and R^2.   [1pt]

Problem   **[10pts]**

We assume that $\mathbf{E}(R)$ is not collinear to 1_d and for $\lambda \in \mathbb{R}$ fixed, we call (P_λ) the problem of finding solutions (B, f) of

$$(P_\lambda) \begin{cases} \mathbf{E}(R - \lambda 1_d - B(f - \lambda)) = 0 \\ \mathbf{Cov}(R - \lambda 1_d - B(f - \lambda), f) = 0 \end{cases}$$

where B is a vector of $\mathbb{R}^d$ and f is a non-constant random variable taking values in $\mathbb{R}$.

(1) Show that:
(B, f) is a solution of (P_λ) $\implies$ $\mathbf{E}(f) \neq \lambda$ and $\frac{\mathbf{Cov}(R, f)}{\mathbf{Var}(f)}(\mathbf{E}(f) - \lambda) = M - \lambda 1_d$.   [2pt]

The result of question (1) can be admitted for the rest of the problem.

For $x \in \mathbb{R}^d$ we define $f_x = x'R$ (if π is a portfolio allocation we then have $f_\pi = \pi'R = R_\pi$ and we use the notations indifferently).

(2) Show that:
$M - \lambda 1_d = \frac{\mathbf{Cov}(R, f_x)}{\mathbf{Var}(f_x)}(\mathbf{E}(f_x) - \lambda) \implies x$ and $\Sigma^{-1}(M - \lambda 1_d)$ are collinear.[1pt]

From now on until the end we assume that $\lambda \neq \frac{b}{a}$.

(3)
(a) If π_λ is an investment portfolio with no risk-free asset allocation and collinear to $\Sigma^{-1}(M - \lambda 1_d)$, show that $\pi_\lambda = \frac{\Sigma^{-1}(M - \lambda 1_d)}{b - \lambda a}$.   [0.5pt]

(b) Express $\mathbf{E}(R_{\pi_\lambda}) - \lambda$ as a function of $a, b, \lambda, M, 1_d$ and $\| \cdot \|_{\Sigma^{-1}}$.   [1pt]

(c) Express $\mathbf{Var}(R_{\pi_\lambda})$ as a function of $a, b, \lambda, M, 1_d$ and $\| \cdot \|_{\Sigma^{-1}}$.   [0.5pt]

(d) Show that $\frac{\mathbf{Cov}(R, R_{\pi_\lambda})}{\mathbf{Var}(R_{\pi_\lambda})}(\mathbf{E}(R_{\pi_\lambda}) - \lambda) = M - \lambda 1_d$.   [1pt]

(e) Assuming that $(B_\lambda, R_{\pi_\lambda})$ is a solution of (P_λ) calculate B_λ as a function of $a, b, \lambda, M, 1_d$ and $\| \cdot \|_{\Sigma^{-1}}$.   [0.5pt]

(f) Verify that $(B_\lambda, R_{\pi_\lambda})$ as defined from 3(a) and 3(e) is a solution of (P_λ).   [1pt]

(4) Show that for solutions (B, f) of (P_λ) we can write $R = \lambda 1_d + B(f - \lambda) + \mathcal{E}$ with $\mathbf{E}(\mathcal{E}) = 0$, $\mathbf{Cov}(f, \mathcal{E}) = 0$ and $\mathrm{Trace}(\mathbf{Var}(\mathcal{E})) < \mathrm{Trace}(\mathbf{Var}(R))$.   [1pt]

(5) Show (using what has been demonstrated so far) that for any investment portfolio of return R_Q we have $\mathbf{E}(R_Q) - \lambda = \frac{\mathbf{Cov}(R_Q, R_{\pi_\lambda})}{\mathbf{Var}(R_{\pi_\lambda})}(\mathbf{E}(R_{\pi_\lambda}) - \lambda).$ **[0.5pt]**

(6) Show (using what has been demonstrated so far) that for any self-financing portfolio of return R_S we have $\mathbf{E}(R_S) = \frac{\mathbf{Cov}(R_S, R_{\pi_\lambda})}{\mathbf{Var}(R_{\pi_\lambda})}(\mathbf{E}(R_{\pi_\lambda}) - \lambda).$ **[0.5pt]**

(7) State (without demonstration) what is the geometric interpretation of the portfolio π_λ.   **[0.5pt]**

10.6.1 Solutions: Exam, January 2018

Exercise 1

(1) $\pi_P^0 = 1 - \pi_P' 1_d$.

(2) $\pi_Q^0 = -\pi_Q' 1_d$.

(3) It is the investment portfolio made of the risky assets of minimum variance.

    (a) $\frac{b}{a}$

    (b) $\frac{1}{a}$.

(4)

    (a) $\alpha = \frac{1}{b - r^0 a}$.

    (b) The Tangent Portfolio.

(5) $R_Q = \frac{\mathbf{Cov}(R_Q, R_T)(R_T - r_0)}{\mathbf{Var}(R_T)} + \epsilon_Q$ with $\mathbf{Cov}(R_T, \epsilon_Q) = 0$ and $\mathbf{E}(\epsilon_Q) = 0$.

(6) $\mathbf{Var}(R) = B\,\mathbf{Var}(F)B' + \mathbf{Var}(\mathcal{E})$.

(7) $\Sigma = \sum\limits_{i=1}^{i=d} \lambda^i u_i u_i'$.

Exercise 2

(1)

    (a) From P_1 and P_3 we get $7\% - r_0 = 0.5(m_T - r_0)$ and $16\% - r_0 = 2(m_T - r_0)$, so $\frac{16\% - r_0}{7\% - r_0} = 4$ and thus $r_0 = 4\%$ and $m_T = 10\%$.
    For portfolio 1: $\sigma(\epsilon_{P_1}) = 0 \Longrightarrow \sigma(R_{P_1}) = \beta_T(P_1)\sigma_T \Longrightarrow \sigma_T = 10\%$.

    (b) $\mathbf{E}(R_{P_2}) - 4\% = 1 \times (10\% - 4\%) \Longrightarrow \mathbf{E}(R_{P_2}) = 10\%$,
    $\mathbf{E}(R_{P_4}) - 4\% = \beta_T(P_4)(10\% - 4\%) \Longrightarrow \beta_T(P_4) = \frac{1}{3}$,
    $\sigma(R_{P_4})^2 = \beta_T(P_4)^2\sigma_T^2 + \sigma(\epsilon_{P_4})^2 \Longrightarrow \sigma(R_{P_4})^2 = \frac{1}{9}(10\%)^2 + (10\%)^2$
    $\Longrightarrow \sigma(R_{P_4}) = \frac{\sqrt{10}}{3}10\%$.

Exercise 3

(1)

    (a) $\pi_S' = (2, -1, -1)$ works and $\mathbf{E}(R_{\pi_S}) = 0$ (this will be the case for all self-financing portfolios in this model which satisfy the AOA conditions).

(b) $\mathbf{Cov}(R_\pi, F) = \pi' B \mathbf{Var}(F)$ so $\mathbf{Cov}(R_\pi, F) = 0 \iff \pi' B = 0$ (as Σ_F is invertible)

but $\pi' B = 0 \implies \pi' \begin{pmatrix} 1 \\ 0 \\ 2 \end{pmatrix} = 0$ and $\pi' \begin{pmatrix} 1 \\ 2 \\ 0 \end{pmatrix} = 0 \implies \pi' 1_3 = 0.$

So, it is not possible to find an investment portfolio such that $\mathbf{Cov}(R_\pi, F) = 0.$

(c) $\mathbf{Var}(R^1) = 3 \times 0.01$, $\mathbf{Var}(R^2) = 5 \times 0.01$.
$\mathbf{Cov}(R^1, R^2) = \mathbf{Cov}(f^1 + f^2 + \epsilon^1, 2f^2 + \epsilon^2) = 2 \times 0.01$ so $\rho = \frac{2}{\sqrt{15}}$.

Problem

(1) $\mathbf{E}(R) - \lambda 1_d \neq 0 \implies \mathbf{E}(f) - \lambda \neq 0.$
$\mathbf{Cov}(R - \lambda 1_d - B(f - \lambda), f) = 0 \iff \mathbf{Cov}(R - Bf, f) = 0$
$\iff \mathbf{Cov}(R, f) - B\mathbf{Cov}(f, f) = 0 \iff B = \frac{\mathbf{Cov}(R,f)}{\mathbf{Var}(f)}.$
$\mathbf{E}(R - \lambda 1_d - B(f - \lambda)) = 0 \iff \mathbf{E}(R) - \lambda 1_d - B(\mathbf{E}(f) - \lambda) = 0$
$\iff B = \frac{M - \lambda 1_d}{\mathbf{E}(f) - \lambda}.$
So, (B, f) solution $\implies \frac{\mathbf{Cov}(R,f)}{\mathbf{Var}(f)} = \frac{M - \lambda 1_d}{\mathbf{E}(f) - \lambda} \implies M - \lambda 1_d = \frac{\mathbf{Cov}(R,f)}{\mathbf{Var}(f)}(\mathbf{E}(f) - \lambda).$

(2) $M - \lambda 1_d = \frac{\mathbf{Cov}(R, f_x)}{\mathbf{Var}(f_x)}(\mathbf{E}(f_x) - \lambda) \iff M - \lambda 1_d = \frac{\mathbf{Cov}(R, x'R)}{\mathbf{Var}(f_x)}(\mathbf{E}(f_x) - \lambda)$
$\iff M - \lambda 1_d = \frac{\Sigma x}{\mathbf{Var}(f_x)}(\mathbf{E}(f_x) - \lambda) \iff x = \Sigma^{-1}(M - \lambda 1_d)\frac{\mathbf{Var}(f_x)}{\mathbf{E}(f_x) - \lambda}.$

(3)

(a) We search for π_λ in the form $\alpha \Sigma^{-1}(M - \lambda 1_d)$.
$1'_d \alpha \Sigma^{-1}(M - \lambda 1_d) = 1 \iff \alpha(b - \lambda a) = 1$ so, $\alpha = \frac{1}{b - \lambda a}$ and $\pi_\lambda = \frac{\Sigma^{-1}(M - \lambda 1_d)}{b - \lambda a}.$

(b) $\mathbf{E}(R_{\pi_\lambda}) = \mathbf{E}(\pi'_\lambda R) - \lambda = \pi'_\lambda M - \lambda \pi'_\lambda 1_d$
$= \pi'_\lambda(M - \lambda 1_d) = \frac{(M - \lambda 1_d)' \Sigma^{-1}(M - \lambda 1_d)}{b - \lambda a} = \frac{\|M - \lambda 1_d\|^2_{\Sigma^{-1}}}{b - \lambda a}.$

(c) $\mathbf{Var}(R_{\pi_\lambda}) = \mathbf{Var}(\pi'_\lambda R) = \frac{1}{(b - \lambda a)^2}(M - \lambda 1_d)' \Sigma^{-1} \Sigma \Sigma^{-1}(M - \lambda 1_d) = \frac{\|M - \lambda 1_d\|^2_{\Sigma^{-1}}}{(b - \lambda a)^2}.$

(d) $\mathbf{Cov}(R, R_{\pi_\lambda}) = \mathbf{Cov}(R, \pi'_\lambda R) = \mathbf{Cov}(R, R)\pi_\lambda = \Sigma \Sigma^{-1}\frac{M - \lambda 1_d}{b - \lambda a} = \frac{M - \lambda 1_d}{b - \lambda a}$, so
$\frac{\mathbf{Cov}(R, R_{\pi_\lambda})}{\mathbf{Var}(R_{\pi_\lambda})}(\mathbf{E}(R_{\pi_\lambda}) - \lambda) = \frac{M - \lambda 1_d}{b - \lambda a}\frac{(b - \lambda a)^2}{\|M - \lambda 1_d\|^2_{\Sigma^{-1}}}\frac{\|M - \lambda 1_d\|^2_{\Sigma^{-1}}}{b - \lambda a} = M - \lambda 1_d.$

(e) According to (1)
$(B_\lambda, R_{\pi_\lambda})$ solution of $(P_\lambda) \implies B_\lambda = \frac{\mathbf{Cov}(R, R_{\pi_\lambda})}{\mathbf{Var}(R_{\pi_\lambda})} \implies B_\lambda = \frac{M - \lambda 1_d}{b - \lambda a}\frac{(b - \lambda a)^2}{\|M - \lambda 1_d\|^2_{\Sigma^{-1}}}$
$\implies B_\lambda = (b - \lambda a)\frac{M - \lambda 1_d}{\|M - \lambda 1_d\|^2_{\Sigma^{-1}}}.$

(f) According to (3b) and (3e)

$$E(R-\lambda 1_d - B_\lambda(R_{\pi_\lambda}-\lambda)) = (M-\lambda 1_d) - (b-\lambda a)\frac{M-\lambda 1_d}{\|M-\lambda 1_d\|_{\Sigma^{-1}}^2}\frac{\|M-\lambda 1_d\|_{\Sigma^{-1}}^2}{(b-\lambda a)} =$$

0. $\mathbf{Cov}(R - \lambda 1_d - B_\lambda(R_{\pi_\lambda} - \lambda), R_{\pi_\lambda}) = 0$ as it is the way B_λ has been defined.

(4) $\mathcal{E} = R - \lambda 1_d - B(f - \lambda)$, so

(B, f) solution of $(P_\lambda) \Longrightarrow \big(E(\mathcal{E}) = 0$ and $\mathbf{Cov}(\mathcal{E}, f) = 0\big)$.

In such a situation, $\mathbf{Var}(R) = \mathbf{Var}(\lambda_0 1_d + B(f - \lambda_0) + \mathcal{E}) = \mathbf{Var}(Bf + \mathcal{E})$

$= \mathbf{Var}(Bf) + \mathbf{Var}(\mathcal{E}) = B'\mathbf{Var}(f)B' + \mathbf{Var}(\mathcal{E})$.

But by hypothesis $\mathbf{Var}(f) > 0$ and $B \neq 0$, so $\text{Trace}(B'\mathbf{Var}(f)B') > 0$ and so, $\text{Trace}(\mathbf{Var}(\mathcal{E})) < \text{Trace}(\mathbf{Var}(R))$.

(5) According to 3(e) $(B_\lambda, R_{\pi_\lambda})$ is a solution of (P_λ) and so according to (1)

$$M - \lambda 1_d = \frac{\mathbf{Cov}(R, R_{\pi_\lambda})}{\mathbf{Var}(R_{\pi_\lambda})}(E(R_{\pi_\lambda}) - \lambda).$$

Therefore, if we call π_Q the allocation of the portfolio Q

$$\pi_Q'(M - \lambda 1_d) = \pi_Q'\frac{\mathbf{Cov}(R, R_{\pi_\lambda})}{\mathbf{Var}(R_{\pi_\lambda})}(E(R_{\pi_\lambda}) - \lambda)$$

$$\Longrightarrow E(R_Q) - \lambda = \frac{\mathbf{Cov}(\pi_Q'R, R_{\pi_\lambda})}{\mathbf{Var}(R_{\pi_\lambda})}(E(R_{\pi_\lambda}) - \lambda)$$

$$\Longrightarrow E(R_Q) - \lambda = \frac{\mathbf{Cov}(R_Q, R_{\pi_\lambda})}{\mathbf{Var}(R_{\pi_\lambda})}(E(R_{\pi_\lambda}) - \lambda).$$

(6) If we call π_S the allocation of the portfolio, the calculation is the same as before, but this time $\pi_S'1_d = 0$, which leads to the new equation.

(7) It is the tangent portfolio obtained by taking $r_0 = \lambda$.

10.7   Midterm Exam, October 2018

Master M1: Mido 29th October 2018 (Midterm Exam: Portfolio Management[7]: Time 2 h)

Notations We consider d risky assets $S_1, S_2, ...S_d$, whose returns over $[0, T]$ satisfy $R^i = m^i + \epsilon^i$. In vector form, we write

$$R = M + \epsilon \text{ with } R = \begin{pmatrix} R^1 \\ \vdots \\ R^d \end{pmatrix}, M = \begin{pmatrix} m^1 \\ \vdots \\ m^d \end{pmatrix} \text{ and } \epsilon = \begin{pmatrix} \epsilon^1 \\ \vdots \\ \epsilon^d \end{pmatrix},$$

---

[7]Pierre Brugière, University Paris 9 Dauphine.

where M is a vector of $\mathbb{R}^d$, ϵ is a Gaussian vector of expectation zero and with invertible matrix of variance-covariance Σ. We assume that there is a risk-free asset S_0 of return r_0.

We denote by $\Pi = \begin{pmatrix} \pi^0 \\ \pi \end{pmatrix}$ an asset allocation where π^0 is the allocation in the risk-free asset and $\pi' = (\pi^1, \cdots, \pi^d)$ the allocation in the risky assets S_i. R_Π is the return of the portfolio Π over $[0, T]$. $\mathbf{E}(R_\Pi)$ is its expectation and $\sigma(R_\Pi)$ its standard deviation. 1_d is the vector of $\mathbb{R}^d$ with all components equal to 1. e_i is the vector of $\mathbb{R}^d$ with all components equal to zero except for the ith component, which equals 1. $a = 1_d' \Sigma^{-1} 1_d$ and $b = 1_d' \Sigma^{-1} M$ and we assume that $r_0 \neq \frac{b}{a}$ and $M \neq r_0 1_d$.

We denote by Φ the cumulative distribution function for a normal law $\mathcal{N}(0, 1)$, so if $Z \sim \mathcal{N}(0, 1)$ then $\forall x \in \mathbb{R}$, $\Phi(x) = P(Z \leq x)$ and Φ^{-1} is its inverse from $]0, 1[$ to $\mathbb{R}$. We denote by ϕ the derivative of Φ, i.e. the density of the distribution function of Z.

Problem (Risk Measures and Capital Allocation)   [20pts]

For any vector $\pi \in \mathbb{R}^d$ we define $\mathbf{RM}_\lambda(\pi) = -\pi'(M - r_0 1_d) + \lambda \sqrt{\pi' \Sigma \pi}$, which is called the Markowitz risk measure of parameter λ for the risk exposure π. $\mathbf{RM}_\lambda(\pi)$ can be interpreted as the capital required for a company to hold the risky positions defined by π.

(1)
   (a) What is the relationship between π^0 and π for an investment portfolio Π of risky allocation π?   [0.5pt]
   (b) Express as a function of r_0, M and π the expected return for an investment portfolio Π of risky allocation π.   [0.5pt]
   (c) Express as a function of π and Σ the standard deviation of the returns for an investment portfolio Π of risky allocation π.   [0.5pt]
   (d) Show that for any $\pi \in \mathbb{R}^d$, $\mathbf{RM}_\lambda(\pi) = -\mathbf{E}(R_\Pi - r_0) + \lambda \sigma(R_\Pi)$, where Π is the investment portfolio of risky allocation π.   [0.5pt]

   For any vector $\pi \in \mathbb{R}^d$ we define the random variable $L_\pi = -\pi'(R - r_0 1_d)$ and for $\alpha \in]0, 1[$ we define $\mathbf{VaR}_\alpha(L_\pi)$ by: $\mathbf{VaR}_\alpha(L_\pi) = \inf\{x, P(L_\pi \leq x) \geq \alpha\}$, which is called the value at risk for the risk exposure π.

(2)
   (a) Express the law of L_π as a function of $\mathbf{E}(R_\Pi)$, r_0 and $\sigma(R_\Pi)$, where Π is the investment portfolio of risky allocation π.   [0.5pt]
   (b) Show that $P(L_\pi \leq \mathbf{VaR}_\alpha(L_\pi)) = \alpha$.   [1pt]
   (c) Express $\mathbf{VaR}_\alpha(L_\pi)$ as a function of $E(R_\Pi)$, r_0, $\sigma(R_\Pi)$, α and Φ.   [1pt]
   (d) For which value of $\lambda(\alpha)$ do we have $\forall \pi \in \mathbb{R}^d$, $\mathbf{RM}_{\lambda(\alpha)}(\pi) = \mathbf{VaR}_\alpha(L_\pi)$?   [0.5pt]

   For any $\pi \in \mathbb{R}^d$ we define the quantity $\mathbf{E}_\alpha(L_\pi) = \mathbf{E}(L_\pi | L_\pi \geq \mathbf{VaR}_\alpha(L_\pi))$, which is called the expected shortfall for the risk exposure π.

(3)

(a) Show that if $a \in \mathbb{R}$ and $Z \sim \mathcal{N}(0, 1)$ then $\mathbf{E}(Z|Z \geq a) = \frac{\phi(a)}{1-\Phi(a)}$. **[0.5pt]**

(b) Show that if $a \in \mathbb{R}$ and $X \sim \mathcal{N}(m, \sigma^2)$ then $\mathbf{E}(X|X \geq a) = m + \frac{\phi(\frac{a-m}{\sigma})}{1-\Phi(\frac{a-m}{\sigma})}\sigma$. **[0.5pt]**

(c) Express $\mathbf{E}_\alpha(L_\pi)$ as a function of $\mathbf{E}(R_\Pi)$, r_0, $\sigma(R_\Pi)$, α, ϕ and Φ. **[1.5pt]**

(d) For which value of $\lambda(\alpha)$ do we have $\forall \pi \in \mathbb{R}^d$, $\mathbf{RM}_{\lambda(\alpha)}(\pi) = \mathbf{E}_\alpha$ (L_π)? **[0.5pt]**

In this section we consider the derivative $\frac{\partial \mathbf{RM}_\lambda}{\partial e_i}(\pi)$ as a row vector representing the derivative, calculated at point π, of $\mathbf{RM}_\lambda(\pi)$ in the direction of vector e_i.

(4)

(a) Show that $\forall \pi \in \mathbb{R}^d$, $\mathbf{RM}_\lambda(\pi) \leq \sum_{i=1}^{d} \mathbf{RM}_\lambda(\pi^i e_i)$. **[1pt]**

(b) Show that $\forall \pi \in \mathbb{R}^d$, $\mathbf{RM}_\lambda(\pi) = \sum_{i=1}^{d} \pi^i \frac{\partial \mathbf{RM}_\lambda}{\partial e_i}(\pi)$. **[1pt]**

(c) How can you interpret the results of 4(a) and 4(b) in terms of capital allocation and diversification effect? **[1pt]**

(5)

(a) Show that $\exists \lambda_0 \in \mathbb{R}$ such that

$$\begin{cases} \inf_{\pi \in \mathbb{R}^d} \mathbf{RM}_\lambda(\pi) = -\infty \text{ if } \lambda < \lambda_0 \text{ and} \\ \inf_{\pi \in \mathbb{R}^d} \mathbf{RM}_\lambda(\pi) = 0 \text{ if } \lambda \geq \lambda_0 \end{cases}$$

and express λ_0 as a function of M, r_0, 1_d and Σ. **[2pt]**

(b) Show that if $\lambda > \lambda_0$ then $\forall \pi \in \mathbb{R}^d \setminus \{0\}$, $\mathbf{RM}_\lambda(\pi) > 0$. **[0.5pt]**

(c) Show that $\exists \pi \in \mathbb{R}^d$ such that $E(R_\Pi - r_0) > 0$. **[0.5pt]**

From now on we assume that $\lambda > \lambda_0$ (as defined in 5(a)) and for any $\pi \in \mathbb{R}^d \setminus \{0\}$ we define the Return on Risk-Adjusted Capital as the quantity $\mathrm{RORAC}_\lambda(\pi) = \frac{-\mathbf{E}(L_\pi)}{\mathbf{RM}_\lambda(\pi)}$ and we call $\mathcal{D} = \{\pi^* \in \mathbb{R}^d \setminus \{0\}, \mathrm{RORAC}_\lambda(\pi^*) = \sup_{\pi \in \mathbb{R}^d} \mathrm{RORAC}_\lambda(\pi)\}$.

(6)

(a) Show that $\pi \in \mathbb{R}^d \setminus \{0\}$ maximises $\mathrm{RORAC}_\lambda(\pi)$ if and only if the investment portfolio Π of risky allocation π maximises the Sharpe Ratio $\frac{\mathbf{E}(R_\Pi - r_0)}{\sigma(R_\Pi)}$. **[1.5pt]**

(b) Using (a) determine $\mathcal{D}$ and $\frac{\mathbf{E}(R_{\Pi^*} - r_0)}{\sigma(R_{\Pi^*})}$ for $\pi^* \in \mathcal{D}$. **[1.5pt]**

(c) Calculate $\mathrm{RORAC}_\lambda(\pi^*)$ for $\pi^* \in \mathcal{D}$ as a function of λ and λ_0. **[1pt]**

(d) For $\pi^* \in \mathcal{D}$ calculate $\frac{-\mathbf{E}[L_{\pi^{*i} e_i}]}{\pi^{*i} \frac{\partial \mathbf{RM}_\lambda}{\partial e_i}(\pi^*)}$ when $m^i \neq r_0$ and $\pi^{*i} \neq 0$ as a function of λ and λ_0. **[2pt]**

10.7.1 Solutions: Midterm Exam October 2018

(1)

   (a) For an investment portfolio the sum of the weights is 1, so $\pi^0 = 1 - \pi'1_d$.

   (b) $\mathbf{E}(R_\Pi) = \pi^0 r_0 + \pi'M = (1 - \pi'1_d)r_0 + \pi'M = r_0 + \pi'(M - r_0 1_d)$.

   (c) $\mathbf{Var}(R_\Pi) = \pi'\Sigma\pi$, so $\sigma(R_\Pi) = \sqrt{\pi'\Sigma\pi}$.

   (d) Trivial from the previous questions.

(2)

   (a) $L_\pi \sim \mathcal{N}\big(-\mathbf{E}(R_\Pi - r_0), \sigma^2(R_\Pi)\big)$.

   (b) For a normal distribution the repartition function is a bijection from $\mathbb{R}$ to $]0, 1[$ so $\exists!\beta$ such that $P(L_\pi \leq \beta) = \alpha$ and it is easy to check that $\mathbf{VaR}_\alpha(L_\pi) = \beta$.

   (c) $P(L_\pi \leq \mathbf{VaR}_\alpha(L_\pi)) = \alpha$

$$\iff P(-\mathbf{E}(R_\Pi - r_0) + \sigma(R_\Pi)Z \leq \mathbf{VaR}_\alpha(L_\pi)) = \alpha$$
$$\iff P(Z \leq \tfrac{1}{\sigma(R_\Pi)}(\mathbf{E}[R_\Pi - r_0] + \mathbf{VaR}_\alpha(L_\pi))) = \alpha$$
$$\iff \Phi(\tfrac{1}{\sigma(R_\Pi)}(\mathbf{E}(R_\Pi - r_0) + \mathbf{VaR}_\alpha(L_\pi))) = \alpha$$
$$\iff \mathbf{VaR}_\alpha(L_\pi) = -\mathbf{E}(R_\Pi - r_0) + \Phi^{-1}(\alpha)\sigma(R_\Pi).$$

   (d) For $\lambda(\alpha) = \Phi^{-1}(\alpha)$.

(3)

   (a) $\mathbf{E}(Z|Z \geq a) = \int\limits_a^{+\infty} z\frac{1}{\sqrt{2\pi}}\exp(-\frac{z^2}{2})\frac{1}{1-\Phi(a)}dz = \frac{1}{1-\Phi(a)}[-\frac{1}{\sqrt{2\pi}}\exp(-\frac{z^2}{2})]_a^{+\infty} = \frac{\phi(a)}{1-\Phi(a)}$.

   (b) $\mathbf{E}(X|X \geq a) = \mathbf{E}(m + \sigma Z|m + \sigma Z \geq a) = m + \sigma\mathbf{E}(Z|Z \geq \frac{a-m}{\sigma}) = m + \frac{\phi(\frac{a-m}{\sigma})}{1-\Phi(\frac{a-m}{\sigma})}\sigma$ from the previous question.

   (c) $L_\pi \sim \mathcal{N}(-\mathbf{E}(R_\Pi - r_0), \sigma^2(R_\Pi))$.

     So, if we define $m = -\mathbf{E}(R_\Pi - r_0)$ and $\sigma = \sigma(R_\Pi)$ we get:

     $\mathbf{E}_\alpha(L_\pi) = m + \frac{\phi(\frac{a-m}{\sigma})}{1-\Phi(\frac{a-m}{\sigma})}\sigma$ with $a = m + \Phi^{-1}(\alpha)\sigma$.

     So, $\mathbf{E}_\alpha(L_\pi) = m + \frac{\phi(\Phi^{-1}(\alpha))}{1-\Phi(\Phi^{-1}(\alpha))}\sigma = m + \frac{\phi(\Phi^{-1}(\alpha))}{1-\alpha}\sigma$

     $= -\mathbf{E}(R_\Pi - r_0) + \frac{\phi(\Phi^{-1}(\alpha))}{1-\alpha}\sigma(R_\Pi)$.

   (d) For $\lambda(\alpha) = \frac{\phi(\Phi^{-1}(\alpha))}{1-\alpha}$.

(4)

   (a) $\sum\limits_{i=1}^d \mathbf{RM}_\lambda(\pi_i e_i)$ is the sum of two terms: the first term $\sum\limits_{i=1}^d (\pi^i e_i)'(M - r_0 1_d) = \pi'(M - r_0 1_d)$ by linearity of the scalar product; the second term $\sum\limits_{i=1}^d \lambda\sqrt{(\pi^i e_i)'\Sigma(\pi^i e_i)} = \lambda\sum\limits_{i=1}^d \|\pi^i e_i\|_\Sigma \geq \lambda\|\sum\limits_{i=1}^d \pi^i e_i\|_\Sigma$ by the triangle inequality applied to the norm $\|\cdot\|_\Sigma$, which proves the result.

(b) $\mathbf{RM}_\lambda(\pi) = -\pi'(M - r_0 1_d) + \lambda \|\pi\|_\Sigma$.

$\mathbf{RM}_\lambda(\pi)$ is positive homogeneous of degree 1 as $\forall t > 0, \mathbf{RM}_\lambda(t\pi) = t\mathbf{RM}_\lambda(\pi)$, so Euler's formula proves the result.

(c) $\mathbf{RM}_\lambda(\pi)$ is the capital required to hold the position π. Equation 4(a) shows the diversification effect and that less capital is required to hold the positions on an aggregated basis than to hold them on a separate basis. Equation 4(b) shows a way to allocate the capital between all the positions after taking into account the diversification effect.

(5)

(a) $\mathbf{RM}_\lambda(\pi) = -\langle \Sigma\pi, M - r_0 1_d \rangle_{\Sigma^{-1}} + \lambda \|\Sigma\pi\|_{\Sigma^{-1}}$.

For $\pi \in \mathbb{R}^d$ we use the decomposition $\Sigma\pi = \beta(M - r_0 1_d) + v$, where v is orthogonal to $M - r_0 1_d$ with respect to $\langle \cdot, \cdot \rangle_{\Sigma^{-1}}$ and $\beta \in \mathbb{R}$, and we get

$$\mathbf{RM}_\lambda(\pi) = -\beta \|M - r_0 1_d\|^2_{\Sigma^{-1}} + \lambda \sqrt{\beta^2 \|M - r_0 1_d\|^2_{\Sigma^{-1}} + \|v\|^2_{\Sigma^{-1}}}$$

then if $\lambda < \|M - r_0 1_d\|_{\Sigma^{-1}}$ by taking $v = 0$ we get $\lim\limits_{\beta \longrightarrow +\infty} \mathbf{RM}_\lambda(\pi) = -\infty$

and if $\lambda \geq \|M - r_0 1_d\|_{\Sigma^{-1}}$ then $\mathbf{RM}_\lambda(\pi) \geq 0$ and reaches its minimum value 0 for $\pi = 0$. So the critical λ is $\lambda_0 = \|M - r_0 1_d\|_{\Sigma^{-1}}$.

(b) The result is clear from the expression

$$\mathbf{RM}_\lambda(\pi) = -\beta \|M - r_0 1_d\|^2_{\Sigma^{-1}} + \lambda \sqrt{\beta^2 \|M - r_0 1_d\|^2_{\Sigma^{-1}} + \|v\|^2_{\Sigma^{-1}}}.$$

(c) $M - r_0 1_d \neq 0 \implies \exists \pi \in \mathbb{R}^d$ such that $\pi'(M - r_0 1_d) \neq 0$ so either $\pi'(M - r_0 1_d) > 0$ or $-\pi'(M - r_0 1_d) > 0$.

(6)

(a) $\arg \max\limits_{\pi \in \mathbb{R}^d \setminus \{0\}} \mathrm{RORAC}_\lambda(\pi) = \arg \max\limits_{\pi \in \mathbb{R}^d \setminus \{0\}} \dfrac{E(R_\Pi - r_0)}{-E(R_\Pi - r_0) + \lambda\sigma(R_\Pi)}$

$= \arg \max\limits_{\pi \in \mathbb{R}^d \setminus \{0\}} \dfrac{\frac{E(R_\Pi - r_0)}{\lambda\sigma(R_\Pi)}}{-\frac{E(R_\Pi - r_0)}{\lambda\sigma(R_\Pi)} + 1}$ so we search for

$$\arg \max\limits_{x(\pi) = \frac{E(R_\Pi - r_0)}{\lambda\sigma(R_\Pi)} < 1} \dfrac{x(\pi)}{1 - x(\pi)}$$

as $\dfrac{E(R_\Pi - r_0)}{\lambda\sigma(R_\Pi)} < 1$.

As we can find π such that $\dfrac{E(R_\Pi - r_0)}{\lambda\sigma(R_\Pi)} > 0$ the maximum will be positive and will be obtained for the positive values of $x(\pi)$ as close as possible to 1, so

$$\arg \max\limits_{\pi \in \mathbb{R}^d \setminus \{0\}} \mathrm{RORAC}_\lambda(\pi) = \arg \max\limits_{\pi \in \mathbb{R}^d \setminus \{0\}} \dfrac{E(R_\Pi - r_0)}{\lambda\sigma(R_\Pi)}.$$

(b) We want to maximize $\frac{\mathbf{E}(R_\Pi - r_0)}{\sigma(R_\Pi)} = \frac{\pi'(M - r_0 1_d)}{\sigma(R_\Pi)} = \frac{\langle \Sigma\pi, M - r_0 1_d \rangle_{\Sigma^{-1}}}{\|\Sigma\pi\|_{\Sigma^{-1}}}$.

If we consider $\pi \in \mathbb{R}^d$ and decompose $\Sigma\pi = \beta(M - r_0 1_d) + v$, where v orthogonal to $M - r_0 1_d$ with respect to $\langle \cdot, \cdot \rangle_{\Sigma^{-1}}$, then $\frac{\langle \Sigma\pi, M - r_0 1_d \rangle_{\Sigma^{-1}}}{\|\Sigma\pi\|_{\Sigma^{-1}}} =$

$\frac{\beta \|M - r_0 1_d\|_{\Sigma^{-1}}}{\sqrt{\beta^2 \|M - r_0 1_d\|_{\Sigma^{-1}}^2 + \|v\|_{\Sigma^{-1}}^2}}$ and the maximum is attained for $v = 0$ and $\beta > 0$

and has value $\|M - r_0 1_d\|_{\Sigma^{-1}}$. So, $\mathcal{D} = \{\beta\Sigma^{-1}(M - r_0 1_d), \beta > 0\}$ and $\forall \pi^* \in \mathcal{D}, \frac{\mathbf{E}(R_{\Pi^*} - r_0)}{\sigma(R_{\Pi^*})} = \|M - r_0 1_d\|_{\Sigma^{-1}}$.

(c) $\mathrm{RORAC}_\lambda(\pi) = \frac{-\mathbf{E}(L_\pi)}{\mathrm{RM}_\lambda(\pi)} = \frac{\pi'(M - r_0 1_d)}{-\pi'(M - r_0 1_d) + \lambda\sqrt{\pi'\Sigma\pi}}$

and $\pi^{*'}(M - r_0 1_d) = \beta \|M - r_0 1_d\|_{\Sigma^{-1}}^2$ and $\pi^{*'}\Sigma\pi^* = \beta^2 \|M - r_0 1_d\|_{\Sigma^{-1}}^2$

so $\mathrm{RORAC}_\lambda(\pi^*) = \frac{1}{\frac{\lambda}{\lambda_0} - 1}$.

(d) $-\mathbf{E}(L_{\pi^{*i} e_i}) = (\pi^{*i} e_i)'(M - r_0 1_d) = \pi^{*i}(m^i - r_0)$ and

$\frac{\partial \mathrm{RM}_\lambda}{\partial \pi}(\pi) = -(M - r_0 1_d)' + \lambda \frac{(\Sigma\pi)'}{\sqrt{\pi'\Sigma\pi}}$.

As $\pi^* = \beta\Sigma^{-1}(M - r_0 1_d)$ with $\beta > 0$ and $\pi^{*'}\Sigma\pi^* = \beta^2\lambda_0^2$ we get

$\frac{\partial \mathrm{RM}_\lambda}{\partial \pi}(\pi^*) = -(M - r_0 1_d)' + \lambda\beta \frac{(M - r_0 1_d)'}{\lambda_0 \beta}$ and

$\pi^{*i} \frac{\partial \mathrm{RM}_\lambda}{\partial e_i}(\pi^*) = -\pi^{*i}(m^i - r_0) + \frac{\lambda}{\lambda_0}\pi^{*i}(m^i - r_0)$,

so $\frac{-\mathbf{E}(L_{\pi^{*i} e_i})}{\pi^{*i} \frac{\partial \mathrm{RM}_\lambda}{\partial e_i}(\pi^*)} = \frac{\pi^{*i}(m^i - r_0)}{\pi^{*i}(m^i - r_0)(-1 + \frac{\lambda}{\lambda_0})} = \frac{1}{-1 + \frac{\lambda}{\lambda_0}}$.

The Lagrangian

A

A.1   Main Results

We recall here the Lagrangian principle for solving optimisation problems under constraints.

We denote by (P) the optimisation problem

$$\begin{cases} \inf_{x \in \mathbb{R}^d} f(x) \\ g_1(x) = 0 \\ \quad \vdots \\ g_k(x) = 0 \end{cases}$$

where $f(\cdot)$ and the $g_i(\cdot)$ are differentiable functions from $\mathbb{R}^d$ into $\mathbb{R}$. We also define $\mathcal{D} = \{x \in \mathbb{R}^d, g_1(x) = 0, \cdots, g_k(x) = 0\}$.

Definition A.1.1 The **Lagrangian** of (P) is a function from $\mathbb{R}^{d+k}$ into $\mathbb{R}$ defined by $\mathcal{L}(x, \lambda_1, \cdots, \lambda_k) = f(x) - \sum_{i=1}^{i=k} \lambda_i g_i(x)$.

We define $\lambda = (\lambda_1, \cdots, \lambda_k)'$ and denote by $\mathcal{L}(x, \lambda)$ the Lagrangian. The following proposition explains how the Lagrangian transforms an optimisation problem under constraints into an optimisation problem without constraints.

Property A.1.1 Solving the problem (P) is equivalent to solving the problem

$$\inf_{x \in \mathbb{R}^d} \sup_{\lambda \in \mathbb{R}^k} L(x, \lambda).$$

© Springer Nature Switzerland AG 2020
P. Brugière, *Quantitative Portfolio Management*, Springer Texts in Business and Economics, https://doi.org/10.1007/978-3-030-37740-3

Proof $x \notin \mathcal{D} \implies \exists i, g^i(x) \neq 0 \implies \sup\limits_{\lambda \in \mathbb{R}^k} \mathcal{L}(x, \lambda) = +\infty$ as the term $-\lambda_i g^i(x)$
has an infinite limit when λ_i varies in $\mathbb{R}$. Alternatively,

$$x \in \mathcal{D} \implies \forall \lambda \in \mathbb{R}^k, \mathcal{L}(x, \lambda) = f(x) \implies \sup\limits_{\lambda \in \mathbb{R}^k} \mathcal{L}(x, \lambda) = f(x).$$

So, $\inf\limits_{x \in \mathcal{D}} f(x) = \inf\limits_{x \in \mathcal{D}} \sup\limits_{\lambda \in \mathbb{R}^k} \mathcal{L}(x, \lambda) = \inf\limits_{x \in \mathbb{R}^k} \sup\limits_{\lambda \in \mathbb{R}^k} \mathcal{L}(x, \lambda).$ □

In this book, the problems considered are quadratic positive definite, under affine constraints which define a domain $\mathcal{D}$ which has a non-empty interior. For this type of problem, the property of strong duality holds for the Lagrangian and therefore,

$$\inf\limits_{x \in \mathbb{R}^k} \sup\limits_{\lambda \in \mathbb{R}^k} \mathcal{L}(x, \lambda) = \sup\limits_{\lambda \in \mathbb{R}^k} \inf\limits_{x \in \mathbb{R}^k} \mathcal{L}(x, \lambda).$$

Now, to solve $\sup\limits_{\lambda \in \mathbb{R}^k} \inf\limits_{x \in \mathbb{R}^k} \mathcal{L}(x, \lambda)$ we start by solving $\inf\limits_{x \in \mathbb{R}^k} \mathcal{L}(x, \lambda)$, and for the
problems we consider in this book the solution is reached and satisfies $\frac{\partial \mathcal{L}}{\partial x}(x, \lambda) = 0$.
By solving this equation in x we find a solution $x^*(\lambda)$ dependent on λ.

The next step is then to solve $\sup\limits_{\lambda \in \mathbb{R}^k} \mathcal{L}(x^*(\lambda), \lambda)$, and in our case the optimum is
reached and satisfies the condition $\frac{\partial \mathcal{L}}{\partial \lambda}(x^*(\lambda), \lambda) = 0$, which is an equation in λ for
which we find a unique solution λ^*.

This being said, the proposition below is useful as it proves that it is not necessary
to solve $\frac{\partial \mathcal{L}}{\partial \lambda}(x^*(\lambda), \lambda) = 0$ but simply to solve $\forall i \in [\![1, k]\!], g(x^*(\lambda)) = 0$.

Property A.1.2

$$\{\lambda \in \mathbb{R}^k, \frac{\partial \mathcal{L}}{\partial \lambda}(x^*(\lambda), \lambda) = 0\} = \{\lambda \in \mathbb{R}^k \text{such that } \forall i \in [\![1, k]\!], g_i(x^*(\lambda)) = 0\}.$$

Proof $\frac{\partial \mathcal{L}}{\partial \lambda}(x^*(\lambda), \lambda) = 0$

$\Longleftrightarrow \forall j \in [\![1, k]\!], \sum\limits_{i=1}^{i=d} \frac{\partial f}{\partial x_i}(x^*(\lambda)) \frac{\partial x_i^*}{\partial \lambda_j}(\lambda) - g_j(x^*(\lambda)) - \sum\limits_{l=1}^{k} \left(\lambda_l \sum\limits_{i=1}^{i=d} \frac{\partial g_l}{\partial x_i}(x^*(\lambda)) \frac{\partial x_i^*}{\partial \lambda_j}(\lambda) \right) = 0$

$\Longleftrightarrow \forall j \in [\![1, k]\!], \sum\limits_{i=1}^{i=d} \frac{\partial f}{\partial x_i}(x^*(\lambda)) \frac{\partial x_i^*}{\partial \lambda_j}(\lambda) - g_j(x^*(\lambda)) - \sum\limits_{i=1}^{i=d} \left(\sum\limits_{l=1}^{k} \lambda_l \frac{\partial g_l}{\partial x_i}(x^*(\lambda)) \frac{\partial x_i^*}{\partial \lambda_j}(\lambda) \right) = 0$

$\Longleftrightarrow \forall j \in [\![1, k]\!], \sum\limits_{i=1}^{i=d} \left(\frac{\partial f}{\partial x_i}(x^*(\lambda)) - \sum\limits_{l=1}^{k} \lambda_l \frac{\partial g_l}{\partial x_i}(x^*(\lambda)) \right) \frac{\partial x_i^*}{\partial \lambda_j}(\lambda) - g_j(x^*(\lambda)) = 0$

but $x^*(\lambda)$ solves $\frac{\partial \mathcal{L}}{\partial x}(x^*(\lambda), \lambda) = 0$ and so for the first term in brackets above,

$$\frac{\partial f}{\partial x_i}(x^*(\lambda)) - \sum_{l=1}^{k} \lambda_l \frac{\partial g_l}{\partial x_i}(x^*(\lambda)) = 0,$$

and so $\frac{\partial \mathcal{L}}{\partial \lambda}(x^*(\lambda), \lambda) = 0 \iff \forall j \in [\![1, k]\!], g_j(x^*(\lambda)) = 0.$ □

So, in practice to solve (P) first we solve in x the equation $\frac{\partial \mathcal{L}}{\partial x}(x, \lambda) = 0$, which gives a solution $x^*(\lambda)$ dependent on λ. Then we solve in λ the equations $g_i(x^*(\lambda)) = 0$, which gives the solution $x^*(\lambda^*)$ for which the optimum is reached. The next section illustrates the method.

A.1.1 Solution of the Markowitz Problem

In the book we solved purely geometrically the problem

$$(P) \begin{cases} \inf_{\pi \in \mathbb{R}^d} \pi' \Sigma \pi \\ \pi' M = m \\ \pi' 1_d = 1 \end{cases} \text{ where } 1_d \text{ is the vector of } \mathbb{R}^d \text{ with components equal to 1}$$

Searching for the solution with the Lagrangian method we get

$$\mathcal{L}(\pi, \lambda) = \pi' \Sigma \pi - \lambda_1 (\pi' M - m) - \lambda_2 (\pi' 1_d - 1)$$

$$\text{So, } \frac{\partial \mathcal{L}}{\partial \pi}(\pi, \lambda) = 0 \iff 2 \Sigma \pi - \lambda_1 M - \lambda_2 1_d = 0$$

$$\iff \pi = \frac{\lambda_1}{2} \Sigma^{-1} M + \frac{\lambda_2}{2} \Sigma^{-1} 1_d.$$

Now using the constraints we get

$$\pi' 1_d = 1 \iff \frac{\lambda_1}{2} b + \frac{\lambda_2}{2} a = 1,$$

so $\pi = \frac{\lambda_1}{2} \Sigma^{-1} M + (\frac{1}{a} - \frac{\lambda_1}{2} \frac{b}{a}) \Sigma^{-1} 1_d = \frac{1}{a} \Sigma^{-1} 1_d + \frac{\lambda_1}{2} \Sigma^{-1} (M - \frac{b}{a} 1_d)$, which is the solution we found in Chap. 4 for the portfolios on the frontier.

The second constraint $\pi' M = m$ determines λ_1 and the portfolio in a unique way as

$$\pi' M = m \iff \frac{1}{a}b + \frac{\lambda_1}{2}\|M - \frac{b}{a}1_d\|_{\Sigma^{-1}}^2 = m \iff \frac{\lambda_1}{2} = \frac{m - \frac{b}{a}}{\|M - \frac{b}{a}1_d\|_{\Sigma^{-1}}^2}.$$

So, the (investment) portfolio solution of (P) is

$$\pi^* = \frac{1}{a}\Sigma^{-1}1_d + \frac{m - \frac{b}{a}}{\|M - \frac{b}{a}1_d\|_{\Sigma^{-1}}^2}\Sigma^{-1}(M - \frac{b}{a}1_d).$$

Parametrisations

B

B.1   Confidence Domain for an Estimator of M

We want to give a parametric representation of the set

$$\mathcal{E}_q = \{u \in \mathbb{R}^2, (M - u)'\Sigma_2^{-1}(M - u) = q\},$$

where M is a vector of $\mathbb{R}^2$ and Σ_2 is a symmetric positive definite matrix of $\mathbb{R}^2$.

Lemma B.1.1 *Let Σ_d be a symmetric positive definite matrix of $\mathbb{R}^d$ and $(u_i)_{i \in [\![1,d]\!]}$ be an orthonormal basis of eigenvectors of Σ_d with eigenvalues $(\lambda_i)_{i \in [\![1,d]\!]}$, then*

(1) the u_i are also eigenvectors for Σ_d^{-1} with eigenvalues $\frac{1}{\lambda_i}$ and
(2) if $i \neq j$, u_i and u_j are also orthogonal with respect to the inner product defined by Σ_d^{-1}.

Proof For (1), $\Sigma_d u_i = \lambda_i u_i \implies u_i = \lambda_i \Sigma_d^{-1} u_i \implies \Sigma_d^{-1} u_i = \frac{1}{\lambda_i} u_i$, which proves the result. For (2), if $i \neq j$ then $u_i' \Sigma_d^{-1} u_j = u_i' \frac{1}{\lambda_j} u_j = 0$, which proves the result. $\qquad\qquad\square$

Proposition B.1.1 *Let Σ_2 be a symmetric positive definite matrix of $\mathbb{R}^2$ and $(u_i)_{i \in [\![1,2]\!]}$ be an orthonormal basis of eigenvectors of Σ_2 with eigenvalues $(\lambda_i)_{i \in [\![1,2]\!]}$, then*

$$\mathcal{E}_q = \{M + \cos(\theta)\sqrt{\lambda_1 q}\, u_1 + \sin(\theta)\sqrt{\lambda_2 q}\, u_2, \theta \in [0, 2\pi]\}.$$

© Springer Nature Switzerland AG 2020
P. Brugière, *Quantitative Portfolio Management*, Springer Texts in Business and Economics, https://doi.org/10.1007/978-3-030-37740-3

Proof If u is a vector of $\mathbb{R}^2$, we can write $u = M + x_1 u_1 + x_2 u_2$ and we get

$$(M - u)' \Sigma_2^{-1} (M - u) = q$$

$$\Longleftrightarrow (x_1 u_1 + x_2 u_2)' \Sigma_2^{-1} (x_1 u_1 + x_2 u_2) = q$$

$$\Longleftrightarrow x_1 u_1' \Sigma_2^{-1} u_1 + x_2 u_2' \Sigma_2^{-1} u_2 = q$$

$$\Longleftrightarrow x_1^2 u_1' \frac{1}{\lambda_1} u_1 + x_2^2 u_2' \frac{1}{\lambda_2} u_2 = q$$

$$\Longleftrightarrow \left(\frac{x_1}{\sqrt{\lambda_1 q}} \right)^2 + \left(\frac{x_2}{\sqrt{\lambda_2 q}} \right)^2 = 1 \tag{B.1.1}$$

Equation (B.1.1) is the equation of an ellipse, and finally we get the parametrisation:

$$u = M + \cos(\theta) \sqrt{\lambda_1 q} u_1 + \sin(\theta) \sqrt{\lambda_2 q} u_2,$$

which proves the result. $\square$

B.2   Confidence Domain for an Observation R_i

Here, we search for a domain $\mathcal{D}$ of probability α, for a normal distribution which has the same mean and variance-covariance matrix as the sample. So, with M and Σ_d equal to the values derived from the data set, we solve for $Z \sim \mathcal{N}(M, \Sigma_d)$ the problem.

$$P(Z \in \mathcal{D}) = \alpha.$$

We consider domains $\mathcal{D}_p$ of the form

$$\mathcal{D}_p = \{ x \in \mathbb{R}^d, (x - M)' \Sigma_d^{-1} (x - M) \le p \}.$$

It is easy to show that,

$$Z \sim \mathcal{N}(M, \Sigma_d) \Longrightarrow (Z - M)' \Sigma_d^{-1} (Z - M) \sim \chi^2(d).$$

Here, $d = 2$ and if we denote by $\Phi_{\chi^2}(2)$ the cumulative distribution function of the law $\chi^2(2)$ we get

$$P(Z \in \mathcal{D}_p) = \alpha \Longleftrightarrow P(\chi^2(2) \le p) = \alpha$$

$$\Longleftrightarrow p = \Phi_{\chi^2(2)}^{-1}(\alpha), \text{ which we denote by } p(\alpha).$$

So, the frontier of $\mathcal{D}_{p(\alpha)}$, according to Proposition B.1.1, is the ellipse

$$\mathcal{E}_{p(\alpha)} = \{M + \cos(\theta)\sqrt{\lambda_1 p(\alpha)}u_1 + \sin(\theta)\sqrt{\lambda_2 p(\alpha)}u_2, \theta \in [0, 2\pi]\} \qquad \text{(B.2.1)}$$

with $p(\alpha) = \Phi^{-1}_{\chi^2(2)}(\alpha)$.

Bibliography

1. Abada, P., Benitob, S., & Lopez, C. (2014). A comprehensive review of value at risk methodologies. *The Spanish Review of Financial Economics, 12*(1), 15–32.
2. Ait-Sahalia, Y., & Xiu, D. (2017). Using principal component analysis to estimate a high dimensional factor model with high-frequency data. *Journal of Econometrics, 201*(2), 384–399.
3. Anderson, T. W. (2003). *An introduction to multivariate statistical analysis* (3rd ed.). New York: Wiley.
4. Ang, N. (2014). *Asset management: A systematic approach to factor investing.* Oxford: Oxford University Press.
5. Arrow, K., & Debreu, G. (1954). Existence of an equilibrium for a competitive economy. *Econometrica, 22*(3), 265–290.
6. Artzner, P., Delbaen, F., Eber, J.-M., & Heath, D. (1999). Coherent measures of risk. *Mathematical Finance, 9*, 203–228.
7. Asness, C. S., Moskowitz, T. J., & Pedersen, L. J. (2013). Value and momentum everywhere. *The Journal of Finance, 68*(3), 929–985.
8. Avramov, C. S., Moskowitz, T. J., & Pedersen, L. J. (2013). Value and momentum everywhere. *The Journal of Finance, 68*(3), 929–985.
9. Avramov, D., & Zhou, G., (2010). Bayesian portfolio analysis. *The Annual Review of Financial Economics, 2*, 25–47.
10. Avramov, D. E. (2019). *Topics in asset pricing.* Lectures Notes. IDC Herzliya College.
11. Axler, S. (2014). *Linear algebra done right* (3rd ed., 340 pp.). Cham: Springer.
12. Bai, J.. & Ng, S. (2013, Sept). Principal components estimation and identification of static factors. *Journal of Econometrics, 176*(1), 18–29.
13. Bank for International Settlement. (2016, January). *Standards: Minimum capital requirements for market risk.* ISBN 978-92-9197-416-0 (online).
14. Bank for International Settlement. (2019, Jan). *Minimum capital requirements for market risk* (136 pp.). Revised Feb 2019. https://www.bis.org/bcbs/publ/d457.pdf.
15. Bank for International Settlement. (2019). *Instructions for Basel III monitoring.* July 2019. https://www.bis.org/bcbs/qis/biiiimplmoninstr_jul19.pdf.
16. Barone-Adesi, G., & Giannopoulos, K. (2000). Non-parametric VaR techniques. Myths and realities. *Economic Notes, 30*, 167–181.
17. Bauder, D., Bodnar, T., Parolya, N., & Schmid, W. (2018). Bayesian mean-variance analysis: Optimal portfolio selection under parameter uncertainty. arXiv:1803.03573v1 [q-fin.ST]. 9 March 2018.
18. Bell, A. R., Brooks, C., & Prokopczuk, M. (2013, January 1). *Handbook of research methods and applications in empirical finance.* Cheltenham: Edward Elgar Publishing.
19. Bera, A., & Jarque, C. (1987). A test for normality of observations and regression residuals. *International Statistical Review, 55*, 163–172.

© Springer Nature Switzerland AG 2020
P. Brugière, *Quantitative Portfolio Management*, Springer Texts in Business and Economics, https://doi.org/10.1007/978-3-030-37740-3

20. Black, F. (1972). Capital market equilibrium with restricted borrowing. *The Journal of Business, 45*(3), 445–455.
21. Black, F., & Litterman, R. (1992). Global optimization. *Financial Analysts Journal, 48*(5), Sep/Oct 1992.
22. Boonen, T. J. (2017). Solvency II solvency capital requirement for life insurance companies based on expected shortfall. *European Actuarial Journal, 7*(2), 405–434.
23. Brugière, P. (2016). *Portfolio management* (p. 115). HAL: https://hal.archives-ouvertes.fr/cel-01327673v2.
24. Cai, Z., Fang, Y., & Tian, D. (2018). Econometric modeling of risk measures: A selective review of the recent literature. Working Paper, October 10, 2018.
25. Chamberlain, G. (1983). Funds, factors and diversification in arbitrage pricing models. *Econometrica, 51*, 1305–1323.
26. Chen, N., & Ingersoll, J. (1983). Exact pricing in linear factor models with infinitely many assets: A note. *Journal of Finance, 38*, 985–988.
27. Choueifaty, Y., & Coignard, Y. (2008). Toward maximum diversification. *Journal of Portfolio Management, 35*(1), 40–51.
28. Choueifaty, Y., & Coignard, Y. (2011). Properties of the Most Diversified Portfolio, Working paper.
29. Clauss, P. (2011). *Gestion de Portefeuille, une approche quantitative*. Paris: Dunod.
30. Cochrane, J. H. (2005). *Asset pricing* (Revised edition, p. 568). Princeton: Princeton University Press.
31. Connor, G., & Korajczyk, R. (1993). A test for the number of factors in an approximate factor model. *The Journal of Finance, 48*, 1263–1291.
32. Dacunha-Castelle, D., & Duflo, M. (1986). *Probability and statistics* (Vol. I). New York: Springer.
33. Dacunha-Castelle, D., & Duflo, M. (1986). *Probability and statistics* (Vol. II). New York: Springer.
34. Darren D., Droettboom M., Firing E., Hunter J., & the Matplotlib Development Team. (2012). https://matplotlib.org/gallery/index.html.
35. Davidson, A. C., & Hinkley, B. V. (1997). *Bootstrap methods and their application*. Cambridge Series in Statistical and Probabilistic Mathematics (592 pp.). Cambridge: Cambridge University Press.
36. Debreu, G. (1983). *Mathematical economics*. Cambridge: Cambridge University Press.
37. Dowd, K. (2005). *Measuring market risk* (2nd ed., 410 pp.). New York: Wiley.
38. Dybvig, P., & Ross, S. (1985). Yes, the APT is testable. *Journal of Finance, 40*, 1173–1188.
39. Eaton, M. L. (2007). *Multivariate statistics: A vector space approach*. IMS Lecture Notes Monograph Series (Vol. 53). https://projecteuclid.org/download/pdf_1/euclid.lnms/1196285114.
40. Efron, B., & Tibshirani, R. J. (1994). *An introduction to the bootstrap* (430 pp.). London: Chapman and Hall. New edition, due date December 18, 2019.
41. Engle, R. F., & Manganelli, S. (2001). Value at risk models in finance. ECB Working Paper, No. 75, August 2001.
42. Fama, E. (1965). The behavior of stock market movements. *Journal of Business, 38*, 1749–1778.
43. Fama, E., & French, K. (1993). Common risk factors in the returns on stocks and bonds. *Journal of Financial Economics, 33*, 3–56.
44. Fama, E. F., & French, K. R. (2004). The capital asset pricing model: Theory and evidence. *Journal of Economic Perspectives, 18*(3), 25–46.
45. Fama, E. F., & French, K. R. (2015). A five-factor asset pricing model. *Journal of Financial Economics, 116*(1), 1–22.
46. Fragkiskos, A. (2014). What a CAIA member should know. *Alternative Investment Analyst Review, Q2 2014, 3*(1), 1–18.
47. Frankfurter, G. M., Phillips, H. E., & Seagle, J. P. (1974). Bias in estimating portfolio alphas and beta scores. *Review of Economics and Statistics, 56*, 412–414.

48. Friedman, J., Hastie T., & Tibshirani, R. (2016). *The elements of statistical learning: Data mining, inference, and prediction* (2nd ed.). Springer Series in Statistics. New York: Springer.
49. Guerard, J. B. (Series Editor). (2009, December 12). *Handbook of portfolio construction: Contemporary applications of Markowitz techniques.* New York: Springer.
50. Hilpisch, Y. (2014). *Python for finance.* Cambridge: O'Reilly.
51. Holden, L. (2008). Some properties of Euler capital allocation. Working paper, Norwegian Computing Center, Blindern, Oslo.
52. Hunter, D. R. (2006). *Statistics 553 asymptotic tools.* Lecture Notes Fall 2006, Chapter 6.
53. Jarrow, R. A., Maksimovic, V., Ziemba, W. T. (Series Editor). (1995). *Handbook in operations research and management science* (Vol. 9). Amsterdam: Finance Elsevier.
54. Jensen, M. C. (1968). The performance of mutual funds in the period 1945–1964. *Journal of Finance, 23,* 389–416.
55. Jobson, J. (1982). A multivariate linear regression test of the arbitrage pricing theory. *Journal of Finance, 37,* 1037–1042.
56. Kharroubi, I., & Ben Tahar, I. (2013). *Gestion de Portefeuilles* (76 pp.). Course Manual University Paris 9 Dauphine.
57. Korkie, B., & Turtle, H. J. (2002). A mean-variance analysis of self-financing portfolio. *Management Science, 48*(3), 427–444.
58. Kotz, M., & Naradajah, S. (2004). *Multivariate t-distributions and their applications.* Cambridge: Cambridge University Press.
59. Lie, H., Li, Y., Fan, X., Zhou, Y., Jin, Z., & Liu, Z. (2012). Approaches to VaR, Document Standford University.
60. Lintner, J. (1965). The valuation of risk assets and the selection of risky investments in stock portfolios and capital budgets. *The Review of Economics and Statistics, 47,* 13–37.
61. Markowitz, H. M. (1952, March). Portfolio selection. *Journal of Finance, 7,* 77–91.
62. Markowitz, H. M. (1959). *Portfolio selection: Efficient diversification of investments.* New York: Wiley.
63. Markowitz, H. M., Peter, T. G., & Sharpe, W. F. (2000). *Mean-variance analysis in portfolio choice and capital markets.* New York: Wiley.
64. McNeil, A. J., Frey, R., & Embrechts, P. (2005). *Quantitative risk management: Concepts, techniques and tools.* Princeton: Princeton University Press.
65. Modigliani, F., & Modigliani, L. (1997). Risk-adjusted performance. *The Journal of Portfolio Management, 23*(2), 45–54.
66. Morgan, J. P., & Reuters. (1996). *RiskMetrics^{TM} - Technical Document* (4th ed.). New York, December 17, 1996.
67. Mossin, J. (1966, October). Equilibrium in a capital asset market. *Econometrica, 34*(4), 768–783.
68. Osmundsen, K. K. (2017). Using Expected Shortfall for Credit Risk Regulation, University of Stavanger Working Papers in Economics and Finance 2017/4.
69. Parzen, E. (1962). On estimation of a probability density function and mode. *Financial Analysts Journal. Annals of Mathematical Statistics, 33,* 1065–1076.
70. Qian, E. E. (2016). *Risk parity fundamentals* (246 pp.). New York: Chapman and Hall/CRC.
71. Rabanel, C. (2019). *Optimisation d'un portefeuille monétaire multidevises sous contrainte de risque et de SCR,* University Paris Dauphine P.S.L. Institut des actuaires.
72. Rencher, A. C. (2003). *Methods of multivariate analysis* (2nd ed.). New York: Wiley.
73. Roll, R. (1977). A critique of the asset pricing theory's tests' part I: On past and potential testability of the theory. *Journal of Financial Economics, 4*(2), 129–176.
74. Roll, R., & Ross, S. (1980). An empirical investigation of the arbitrage pricing theory. *Journal of Finance, 35,* 1073–1103.
75. Roncalli, T. (2013). *Introduction to risk parity and budgeting.* Chapman and Hall/CRC Financial Mathematics Series (440 pp.). Boca Raton: CRC Press.
76. Ross, S. (2005). *First course in probability* (7th ed., p. 576). Harlow: Prentice Hall.
77. Ross, S. A. (1976). The arbitrage theory of capital asset pricing. *Journal of Economic Theory, 13*(3) (December), 341–360.

78. Ruppert, D. (2004). *Statistics and finance, an introduction.* Springer Texts in Statistics. New York: Springer.
79. Sabanes Bove, D., & Held, L. (2014). *Applied statistical inference.* Heidelberg: Springer.
80. Sharpe, W. F. (1964). Capital asset prices: A theory of market equilibrium under conditions of risk. *The Journal of Finance, 19*(3), 425–442.
81. Sharpe, W. F. (1967, March). A linear programming algorithm for mutual fund portfolio selection. *Management Science, 13*(7) Series A, Sciences, 499–510.
82. Sharpe, W. F. (1970). *Portfolio theory and capital markets.* New York: McGraw-Hill.
83. Sharpe, W. F. (1992). Asset allocation: Management style and performance measurement. *Journal of Portfolio Management, 18*(2) (Winter), 29–34.
84. Sharpe, W. F. (1994). The Sharpe ratio. *Journal of Portfolio Management, 21*(1), 49–58.
85. Sheikh, A. (1996). *BARRA's Risk Models.* Copyright ©1996 BARRA, Inc.
86. Steiner, B. (2012). *Key financial market concepts* (2nd ed.). New York: Financial Times/Prentice Hall Ltd.
87. Tasche, D. (1999). Risk contribution and performance measurements. Working Paper, Technische Universitet Munchen.
88. Tasche, D. (2006). Measuring Sectoral Diversification in an Asymptotic Multi-Factor Framework. arXiv: physics/0505142.
89. Tasche, D. (2008). Capital Allocation to Business Units and Sub-Portfolios: The Euler Principle. arxiv.org/pdf/0708.2542.pdf
90. Tobin, J. (1958, February). Liquidity preference as behavior towards risk. *Review of Economic Studies, XXV*(2), 65–86, HB1R4.
91. Treynor, J. L. (1961). Toward a theory of market value of risky assets, mimeo, subsequently published in Korajczyk, Robert A. (1999). *Asset pricing and portfolios performance: Models, strategy and performance metrics.* London: Risk Books.
92. Von Neumann, J., & Morgenstern, O. (1947). *Theory of games and economic behavior.* Princeton: Princeton University Press.
93. Von Rosend, D. (1988). Moments for the inverted Wishart Distribution. *Scandinavian Journal of Statistics, 15*(2), 97–109.
94. Wang, J., & Zivot, E. (2005, December). *Modeling financial time series with S-PLUS* (2nd ed.). New York: Springer.

Index

© Springer Nature Switzerland AG 2020
P. Brugière, *Quantitative Portfolio Management*, Springer Texts in Business
and Economics, https://doi.org/10.1007/978-3-030-37740-3

Printed in the United States
by Baker & Taylor Publisher Services